SolidWorks CAM 2023
Black Book

By
Gaurav Verma
Matt Weber
(CADCAMCAE Works)

ISBN # 978-1-77459-053-9

NOTICE TO THE READER

Publisher does not warrant or guarantee any of the products described in the text or perform any independent analysis in connection with any of the product information contained in the text. Publisher does not assume, and expressly disclaims, any obligation to obtain and include information other than that provided to it by the manufacturer.

The reader is expressly warned to consider and adopt all safety precautions that might be indicated by the activities herein and to avoid all potential hazards. By following the instructions contained herein, the reader willingly assumes all risks in connection with such instructions.

The Publisher makes no representation or warranties of any kind, including but not limited to, the warranties of fitness for a particular purpose or merchantability, nor are any such representations implied with respect to the material set forth herein, and the publisher takes no responsibility with respect to such material. The publisher shall not be liable for any special, consequential, or exemplary damages resulting, in whole or part, from the reader's use of, or reliance upon, this material.

DEDICATION

To teachers, who make it possible to disseminate knowledge
to enlighten the young and curious minds
of our future generations

To students, who are the future of the world

THANKS

To my friends and colleagues

To my family for their love and support

Training and Consultant Services

At CADCAMCAE Works, we provides effective and affordable one to one and group online training on various software packages in Computer Aided Design(CAD), Computer Aided Manufacturing(CAM), Computer Aided Engineering (CAE), Computer programming languages(C/C++, Java, .NET, Android, Javascript, HTML and so on). The training is delivered through remote access to your system and voice chat via Internet at any time, any place, and at any pace to individuals, groups, students of colleges/universities, and CAD/CAM/CAE training centers. The main features of this program are:

Training as per your need

Highly experienced Engineers and Technician conduct the classes on the software applications used in the industries. The methodology adopted to teach the software is totally practical based, so that the learner can adapt to the design and development industries in almost no time. The efforts are to make the training process cost effective and time saving while you have the comfort of your time and place, thereby relieving you from the hassles of traveling to training centers or rearranging your time table.

Software Packages on which we provide basic and advanced training are:

CAD/CAM/CAE: CATIA, Creo Parametric, Creo Direct, SolidWorks, Autodesk Inventor, Solid Edge, UG NX, AutoCAD, AutoCAD LT, EdgeCAM, MasterCAM, SolidCAM, DelCAM, BOBCAM, UG NX Manufacturing, UG Mold Wizard, UG Progressive Die, UG Die Design, SolidWorks Mold, Creo Manufacturing, Creo Expert Machinist, NX Nastran, Hypermesh, SolidWorks Simulation, Autodesk Simulation Mechanical, Creo Simulate, Gambit, ANSYS and many others.

Computer Programming Languages: C++, VB.NET, HTML, Android, Javascript and so on.

Game Designing: Unity.

Civil Engineering: AutoCAD MEP, Revit Structure, Revit Architecture, AutoCAD Map 3D and so on.

We also provide consultant services for Design and development on the above mentioned software packages

For more information you can mail us at:
cadcamcaeworks@gmail.com

TABLE OF CONTENTS

Chapter 2 : Setting Up Workpiece and Machine

Chapter 3 : Milling Setup and Features

Chapter 4 : Milling Operations

Chapter 5 : Toolpath Generation and Processing

Chapter 8 : Practical and Practice

Page left blank intentionally

Preface

SOLIDWORKS® CAM - powered by CAMWorks - uses rules-based technology that enables you to integrate design and manufacturing in one application, connecting design and manufacturing teams through a common software tool and 3D model. SOLIDWORKS CAM is an add-on to all versions of SOLIDWORKS CAD that lets you prepare your designs for manufacturability earlier in the development cycle. Manufacturing tasks that had to wait until a design was complete can now be performed concurrently with the design process. SOLIDWORKS CAM is available free with educational package of this software for learning.

The **SolidWorks CAM 2023 Black Book** is the 4th edition of our series on SolidWorks CAM. The book is written to help professionals as well as learners get familiar with functionality of the software. The book follows a step by step methodology. In this book, we have tried to give real-world examples with real challenges in manufacturing design. We have tried to reduce the gap between university use and industrial use of SolidWorks CAM. The book covers almost all the information required by a learner to master SolidWorks CAM. Some of the salient features of this book are :

In-Depth explanation of concepts
Every new topic of this book starts with the explanation of the basic concepts. In this way, the user becomes capable of relating the things with real world.

Topics Covered
Every chapter starts with a list of topics being covered in that chapter. In this way, the user can easy find the topic of his/her interest easily.

Instruction through illustration
The instructions to perform any action are provided by maximum number of illustrations so that the user can perform the actions discussed in the book easily and effectively. There are about 400 small and large illustrations that make the learning process effective.

Tutorial point of view

At the end of concept's explanation, the tutorial make the understanding of users firm and long lasting. Most of the tools in this book are discussed in the form of tutorials.

Project

Projects and exercises are provided to students for practicing.

For Faculty

If you are a faculty member, then you can ask for video tutorials on any of the topic, exercise, tutorial, or concept. As faculty, you can register on our website to get electronic desk copies of our latest books, self-assessment, and solution of practical. Faculty resources are available in the **Faculty Member** page of our website (**www.cadcamcaeworks.com**) once you login. Note that faculty registration approval is manual and it may take two days for approval before you can access the faculty website.

Formatting Conventions Used in the Text

All the key terms like name of button, tool, drop-down etc. are kept bold.

Free Resources

Link to the resources used in this book are provided to the users via email. To get the resources, mail us at ***cadcamcaeworks@gmail.com*** with your contact information. With your contact record with us, you will be provided latest updates and informations regarding various technologies. The format to write us mail for resources is as follows:

Subject of E-mail as ***Application for resources of _____ book***.
Also, given your information like
Name:
Course pursuing/Profession:
E-mail ID:

Note: We respect your privacy and value it. If you do not want to give your personal informations then you can ask for resources without giving your information.

About Authors

The author of this book, Gaurav Verma, has authored and assisted in more than 18 titles in CAD/CAM/CAE which are already available in market. He has authored **AutoCAD Electrical Black Books** which are available in both **English** and **Russian** language. He has also written **Creo Manufacturing 9.0 Black Book** which covers Expert Machinist module of Creo Parametric. He has provided consultant services to many industries in US, Greece, Canada, and UK. He has assisted in preparing many Government aided skill development programs. He has been speaker for Autodesk University, Russia 2014. He has assisted in preparing AutoCAD Electrical course for Autodesk Design Academy. He has worked on Sheetmetal, Forging, Machining, and Casting areas of Design and Development department.

The author of this book, Matt Weber, has authored many books on CAD/CAM/CAE available already in market. **SolidWorks Simulation Black Books** and **SolidWorks Flow Simulation Black Books** are one of the most selling books in SolidWorks Simulation field. The author has hands on experience on many popular CAD/CAM/CAE packages. If you have any query/doubt in any CAD/CAM/CAE package, then you can contact the author by writing at cadcamcaeworks@gmail.com

For Any query or suggestion

If you have any query or suggestion, please let us know by mailing us on *cadcamcaeworks@gmail.com*. Your valuable constructive suggestions will be incorporated in our books.

Page left blank intentionally

Chapter 1

Introduction

Topics Covered

The major topics covered in this chapter are:

- *Introduction to manufacturing.*
- *Types of Machines.*
- *Applications of CAM.*
- *Activating SolidWorks CAM Add-in.*
- *General Approach in CAM.*
- *Defining Machine.*
- *Setting Coordinate System.*
- *Defining or Editing Stock.*
- *Walkthrough of SolidWorks CAM.*

INTRODUCTION TO MANUFACTURING

Manufacturing is the process of creating a useful product by using a machine, a process, or both. For manufacturing a product, there are some steps to be followed:

- Generating Layout of final product.
- Selection of Raw material/Work piece; selection of raw material depends on the application of the product.
- Forging, Casting, or any other pre-machining method for creating outlines for final shape.
- Roughing Processes.
- Finishing Processes.
- Quality Control.

As the "Generating Layout of final product" is above all the steps, it is the most important step. One should be very clear about the final product because all the other steps are totally dependent on the first step. The layout of final product can be a drawing or a model created by using any modeling software like SolidWorks. (Refer to our another title **SolidWorks 2023 Black Book** for concepts of modeling)

The next step is "Selection of Raw material/Workpiece". This step is solely dependent on the first step. Our final product defines what should be the raw material and the workpiece. Here workpiece is the piece of raw material to be used for the next step or process.

The next step is "Forging, Casting or any other process for creating outline of the final shape". The outline created for the final shape is also called "Blank" in industries. In this step, various machines like Press, Cutter, or Moulding machines are used for creating the blank. In some cases of Casting, there is no requirement of machining processes. For example, in case of Investment casting most of the time there is no requirement of machining process. Machining processes can be divided further into two processes:

- Roughing Processes
- Finishing Processes

These processes are the main discussion area of this book. An introduction to these processes is given next.

Roughing Process

Roughing is the starting of machining process. Generally in roughing process, large amount of stock material is removed as compared to finishing process. In a roughing process, the quantity of material removed from the workpiece is more important than the quality of the machining. Generally, there are no close tolerances for roughing processes. So, these processes are relatively cheaper than the finishing processes. In manufacturing industries, there are three principle machining processes called Turning, Milling, and Drilling. In case of roughing process, there can be turning, milling, drilling, combination of any two, or all the processes. Apart from these principle machining processes, there are various other processes like shaping, planing, broaching, reaming, and so on. But these processes are used in special cases.

Finishing Process

Finishing process can include all the machining processes discussed in case of roughing processes but in close tolerances. Also, the quality of machining at required accuracy level is very important for finishing. There are a few more machining processes like Electric Discharge Machining (EDM), Laser Beam Machining, Electrochemical Machining, and so on. These processes are called unconventional machining processes because of their cutting method. In unconventional machining processes, the tool life is much higher than the conventional machining. Different Machines used for machining processes are discussed next.

TYPES OF MACHINES

There are various types of machines for different type of machining process. For example- for turning process, there are machines like conventional lathe and CNC Turner. Similarly for milling process, there are machines called Milling machine, VMC, or HMC. Some of the machines are discussed next with details of their functioning.

Turning Machines

Turning machine is a category of machines used for turning process. In this machine, the workpiece is held in a chuck (collet in case of small workpieces). This chuck revolves at a defined rotational speed. Note that the workpiece can revolve in either CW(Clockwise) or CCW(Counter-Clockwise) direction but cannot translate in any direction. The cutting tool used for removing material can translate in X and Y directions. The most basic type of turning machine is a lathe. But now a days, lathes are being replaced by CNC Turning machines, which are faster and more accurate then the traditional lathes. The CNC Turning machines are controlled by numeric codes. These codes are interpreted by machine controller attached in the machine and then the controller commands various sections of the machine to do a specific job. The basic operations that can be done on turning machines are:

- Taper turning
- Spherical generation
- Facing
- Grooving
- Parting (in few cases)
- Drilling
- Boring
- Reaming
- Threading

Milling Machines

Milling machine is a category of machines used for removing material by using a perpendicular tool relative to the workpiece. In this type of machine, workpiece is held on a bed with the help of fixtures. The tool rotates at a defined speed. This tool can move in X, Y, and Z directions. In some machines, the bed can also translate and rotate like in Turret milling machines, 5-axis machines, and so on. Milling machines are of two types; horizontal milling machine and vertical milling machine. In Horizontal milling machine, the tool is aligned with the horizontal axis (X-axis). In Vertical milling machine, the tool is aligned with the vertical axis (Z-axis). The Vertical milling machine is generally used for complex cutting processed like contouring, engraving, embossing and so on. The Horizontal milling machines are used for cutting slots, grooves, gear teeth, and so on. In some Horizontal milling machines, table can moved up-down by motor mechanism or power system. By using the synchronization of table movement with the rotation of rotary fixture, we can also create spiral features. The tools used in both type of milling machines have cutting edge on the sides as well as at the tip.

Drilling Machines

Drilling machine is a category of machines used for creating holes in the workpiece. In Drilling machine, the tool (drill bit) is fixed in a tool holder and the tool can move up-down. The workpiece is fixed on the bed. The tool goes down, mechanism or by manual handle, penetrating through the workpiece. There are various types of Drilling machine available like drill presses, cordless drills, pistol grip drills and so on.

Shaper

Shaper is a category of machines, which is used to cut material in a linear motion. Shaper has a single point cutting tool, which goes back-forth to create linear cut in the workpiece. This type of machine is used to create flat surface of the workpiece. You can create dovetail slots, splines, key slots, and so on by using this machine. In some operations, this machine can be an alternative for EDM.

Planer

Planer is a category of machines similar to Shaper. The only difference is that, in case of Planer machine, the workpiece reciprocates and the tool is fixed.

There are various other special purpose machines (SPMs), which are used for some uncommon requirements. The machines discussed above are conventional machines. The unconventional machines are discussed next.

Electric Discharge Machine

Electric Discharge Machine is a category of machines used for creating desired shapes on the workpiece with the help of electric discharges. In this type of machines, the tool and the workpiece act as electrodes and a dielectric fluid is passed between them. The workpiece is fixed in the bed and tool can move in X, Y, and Z direction. During the machining process, the tool is brought near to the workpiece. Due to this, a spark is generated between them. This spark causes the material on the workpiece to melt and get separated from the workpiece. This separated material is drained with the help of dielectric fluid. There are two types of EDMs which are listed next.

Wire-cut EDM

In this type of EDM, a brass wire is commonly used to cut the material from the workpiece. This wire is held in upper and lower diamond shaped guides. It is constantly fed from a bundle. In this machine, the material is removed by generating sparks between tool and workpiece. A Wire-cut EDM can be used for a plate having thickness up to 300 mm.

Sinker EDM

In this type of EDM, a metal electrode is used to cut the material from the workpiece. The tool and the workpiece are submerged in the dielectric fluid. Power supply is connected to both the tool and the workpiece. When tool is brought near the workpiece, sparks are generated randomly on their surfaces. Such sparks gradually create impression of tool on the workpiece.

Electro Chemical Machine

Electro Chemical Machine is a category of machines used for creating desired shape by using the chemical electrolyte. This machining works on the principles of chemical reactions.

Laser Beam Machine

Laser Beam Machine is a category of machines that uses a beam of highly coherent light. This type of light is called laser. A laser can output a power of up to 100MW in an area of 1 square mm. A laser beam machine can be used to create accurate holes or shapes on a material like silicon, graphite, diamond, and so on.

The machines discussed till now are the major machines used in industries. Some of these machines can be controlled by numeric codes and are called NC machines. NC Machines and their working is discussed next.

NC MACHINES

An NC Machine is a manufacturing tool that removes material by following a predefines command set. An NC Machine can be a milling machine or it can be a turning center. NC stands for Numerical Control so, these machines are controlled by numeric codes. These codes are dependent on the controller installed in the machines. There are various controllers available in the market like Fanuc controller, Siemens controller, Heidenhain controller, and so on. The numeric codes change according to the controller used in the machine. These numeric codes are compiled in the form of a program, which is fed in the machine controller via a storage media. The numeric codes are generally in the form of G-codes and M-codes. For understanding purpose, some of the G-codes and M-codes are discussed next with their functions for a Fanuc controller.

Code		Function
G00	-	Rapid movement of tool.
G01	-	Linear movement while creating cut.
G02	-	Clockwise circular cut.
G03	-	Counter-clockwise circular cut.
G20	-	Starts inch mode.
G21	-	Starts mm mode.
G96	-	Provides constant surface speed.
G97	-	Constant RPM.
G98	-	Feed per minute
G99	-	Feed per revolution

M00 -   Program stop
M02 -   End of program
M03 -   Spindle rotation Clockwise.
M04 -   Spindle rotation Counter Clockwise.
M05 -   Spindle stop
M08 -   Coolant on
M09 -   Coolant off
M98 -   Subprogram call
M99 -   Subprogram exit

These codes as well as the other codes will be discussed in the subsequent chapters according to their applications.

As there is a long list of codes which are required in NC programs to make machine cut workpiece in desired size and shape, it becomes a tedious job to create programs manually for each operation. Moreover, it take much time to create a program for small operations on a milling machine. To solve this problem and to reduce the human error, Computer Aided Manufacturing (CAM) was introduced. Various applications of CAM are discussed next.

APPLICATIONS OF COMPUTER AIDED MANUFACTURING

Computer Aided Manufacturing (CAM) is a technology which can be used to enhance the manufacturing process. In this technology, the machines are controlled by a workstation. This workstation can serve more than one machines at a time. Using CAM, you can create and manage the programs being fed in the workstation. Some of the applications of CAM are discussed next.

1. CAM with the combination of CAD can be used to create complex shapes by machining in less time.
2. CAM can be used to manage more than one machines at the same time with less human power.
3. CAM is used to automate the manufacturing process.
4. CAM is used to generate NC programs for various types of NC machines.
5. 5-Axis Machining can be performed easily with CAM

CAM is generally the next step after CAD (Computer Aided Designing). Sometimes CAE (Computer Aided Engineering) is also required before CAM. There are various software companies that provide the CAM software solutions. SolidWorks is one of the most popular 3D CAD software. SOLIDWORKS CAM is an add-in designed to automate manufacturing programming for 3D data created in SOLIDWORKS, marking another step toward manufacturing information arriving in the shop without drawings. This add-in is powered by long term SolidWorks CAM partner, CAMWorks. SolidWorks CAM provides smart manufacturing capability which includes Model-Based Definition (MBD), Costing, and Inspection information for generating manufacturing data.

Applications of SolidWorks CAM

Following are some of the applications of SolidWorks CAM:

- It enables users to program in either an assembly or a part environment.
- It can interpret surface finishes and tolerances to optimize the best routes for manufacturing a part.
- It automatically applies standard manufacturing strategies to increase efficiency and uniformity.
- It performs automated quoting and analyzes against traditional methods to account for every characteristic of the part in advance.
- It makes automatic adjustments of machining strategies based on tolerance specifications and model-based definition.
- Its Automatic Feature Recognition gives users automatically generated machine programming for prismatic parts by referencing programming standards.
- It has 2.5-axis functionality with part and assembly machining.

ACTIVATING SOLIDWORKS CAM ADD-IN

- Start SolidWorks by using the Start menu or icon on Desktop.
- Click on the down button next to **Options** button in the Quick Access toolbar; refer to Figure-1 and click on the **Add-Ins** button or click on the **Tools > Add-Ins** button from the menu. The **Add-Ins** box will be displayed; refer to Figure-2.

Figure-1. Quick Access toolbar

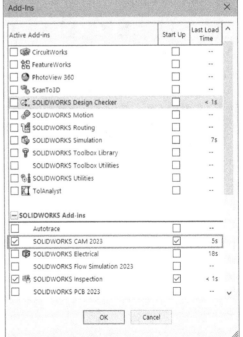

Figure-2. Add-Ins box

- Scroll-down in the box and select the check box before the **SOLIDWORKS CAM 2023** option to activate the application. If you want SolidWorks CAM to start with SolidWorks each time then also select the check box after the **SOLIDWORKS CAM 2023** option as well in this box.
- Click on the **OK** button from the box.

By default, the **SolidWorks CAM** Add-In is active in SolidWorks and the tools related to CAM are available in **SOLIDWORKS CAM CommandManager**; refer to Figure-3.

Figure-3. SOLIDWORKS CAM CommandManager

Note that various trees are available for SolidWorks CAM in the left area of the application window; refer to Figure-4.

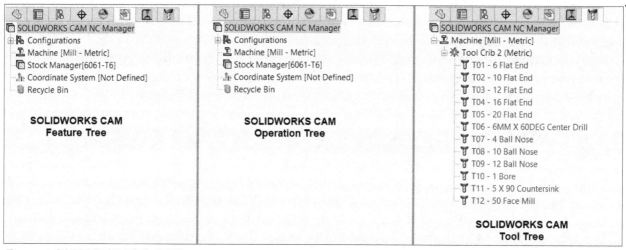

Figure-4. SOLIDWORKS CAM Tree

BASIC APPROACH FOR CAM

Whether you use the stand-alone program like Mastercam or the integrated one in SolidWorks like SolidWorks CAM, the approach for creating NC programs is same. First, you need to import or create the CAD model of the product. Then, create stock of material (workpiece) from which the product will be manufactured after machining. Apply settings related to machine. Apply parameters related to tools. Create the tool paths for operations to be performed on the machine. Simulate the machining process and check whether it is as per the requirement. Generate the output of the machining which is NC codes. Refer to Figure-5.

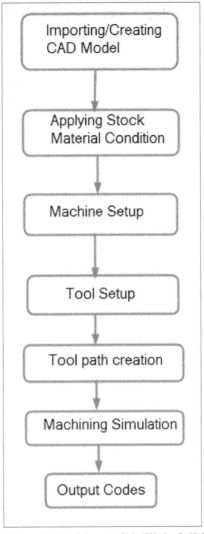

Figure-5. Workflow in SolidWorks CAM

The basic overview of SolidWorks CAM is discussed next.

DEFINING MACHINE

The **Define Machine** tool is used to define the type of machine and other related parameters for creating machine setup as per the job. The procedure to define machine is given next.

- After opening the part file, right-click on the **Machine [Mill-Metric]** option from the **SOLIDWORKS CAM Feature Tree** and select **Edit Definition** option; refer to Figure-6. The **Machine** dialog box will be displayed; refer to Figure-7.

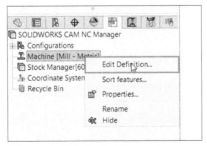

Figure-6. Edit Definition option

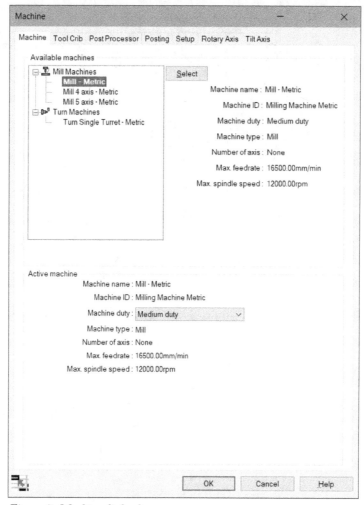

Figure-7. Machine dialog box

- Select desired machine from the list of machine and set the parameters related to machine.

Setting Tool Crib

- Click on the **Tool Crib** tab from the dialog box. The dialog box will be displayed as shown in Figure-8.
- By default, tools for performing various operations are available in the tool crib. You can add or remove the tools based on your machine's tool cassette.

Adding Tools

- To add a tool, click on the **Add Tool** button from the dialog box. The **Tool Select Filter** dialog box will be displayed; refer to Figure-9.

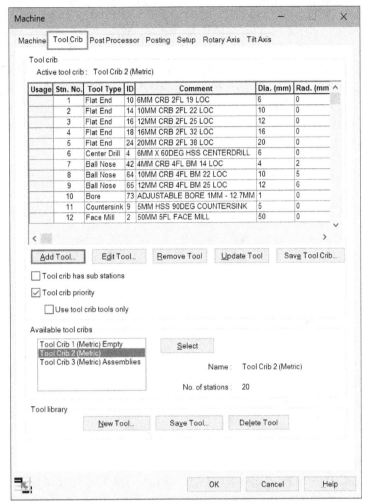

Figure-8. Tool Crib tab in Machine dialog box

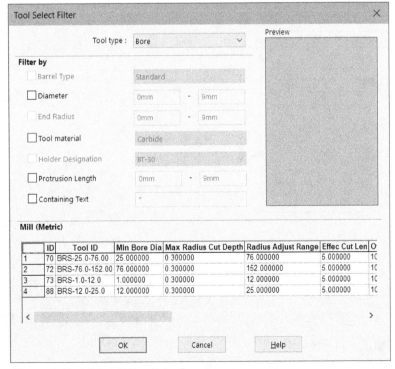

Figure-9. Tool Select Filter dialog box

- Click in the **Tool type** drop-down and select desired type of tool to be added in the crib. Like if you want to use Ball Nose cutting tool then select the **Ball Nose** option from the drop-down.
- Set desired filters for the tool and select desired tool to be added.
- You can also search a tool by specifying keywords of tool name. To do so, select the **Containing Text** check box, type the name of desired tool in respective edit box, and press **TAB**. The list of tools containing keyword will be displayed; refer to Figure-10.

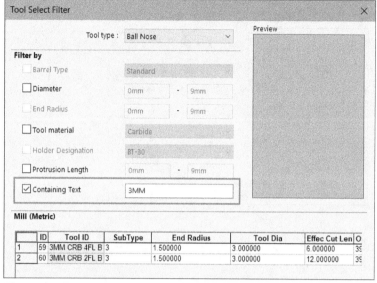

Figure-10. Tool filter by text

- Select desired tool from the list and click on the **OK** button from the dialog box. The tool will be added in the list and the preview of tool will be displayed; refer to Figure-11.

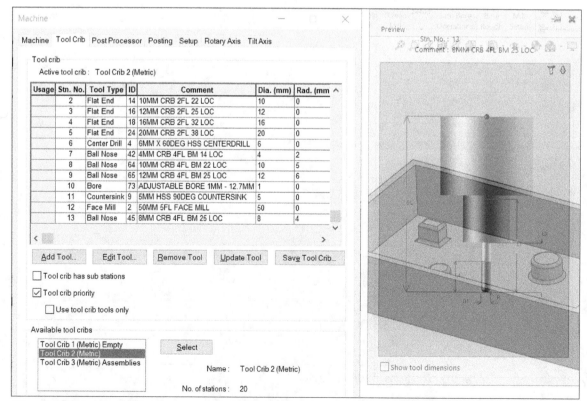

Figure-11. Preview of the tools

Editing Tool

- Select the tool you want to edit from the list and click on the **Edit Tool** button. The **Edit Tool Parameters** dialog box will be displayed; refer to Figure-12.
- Set desired parameters in the dialog box. Note that you can set the tool holder and tool station as well for your tool.
- After setting the tool, click on the **OK** button. (You will learn more about tool parameters later in the book.)

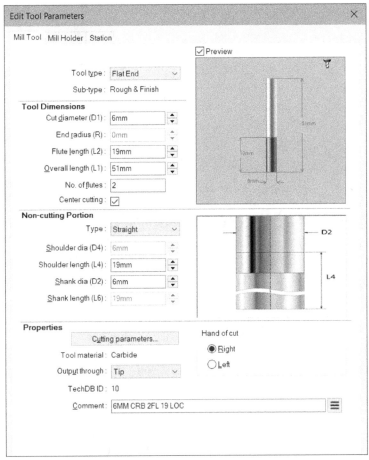

Figure-12. Edit Tool Parameters dialog box

Removing Tool from the Crib

- Select the tool that you want to remove from crib and click on the **Remove Tool** button from the dialog box. The tool will be removed.

Updating tool in Database

Based on the parameters defined by you in the dialog box, you can update the definitions of the tools in the central database of SolidWorks. To do so, after editing tools; click on the **Update Tool** button from the dialog box.

Saving Tool Crib

After creating desired tool crib, you can save it for later use. The procedure is given next.

- Click on the **Save Tool Crib** button from the dialog box. The **Save to Database** dialog box will be displayed; refer to Figure-13.

*   Set desired name and parameters, and then click on the **Save** button.

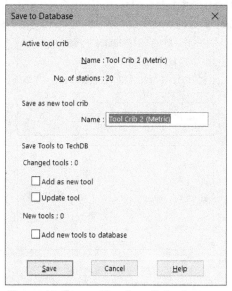

Figure-13. Save to Database dialog box

Creating and Saving New Tool in Library

*   Click on the **New Tool** button from the **Tool library** area of the dialog box. The **New Tool** dialog box will be displayed similar to **Edit Tool Parameters** dialog box; refer to Figure-14.

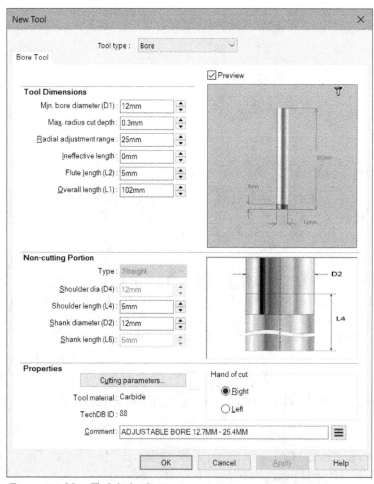

Figure-14. New Tool dialog box

- Set the parameters as discussed earlier and click on the **OK** button to create the tool. The tool will be created and added to the library as well as tool crib.
- If a tool is not added in the library, then select it from the crib table and click on the **Save Tool** button from the **Tool library** area.

Setting Postprocessor Parameters

The post processor is used to translate the codes generated by CAM program to machine readable codes based on machine controller. The options for post processor are available in the **Post Processor** tab of the **Machine** dialog box; refer to Figure-15.

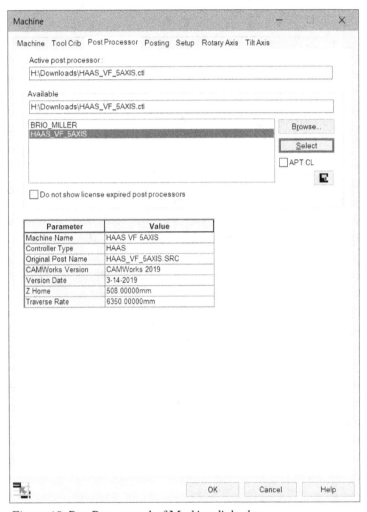

Figure-15. Post Processor tab of Machine dialog box

- Select desired post processor from the list and click on the **Select** button.
- If you have an updated post processor then click on the **Browse** button. The **Open** dialog box will be displayed.
- Select desired post processor and click on the **Open** button.

Posting Options

The options related to posting coolant offset are available in the **Posting** tab; refer to Figure-16.

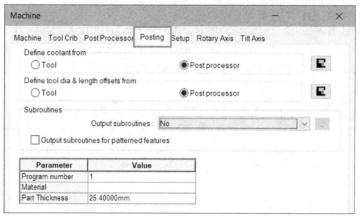

Figure-16. Posting tab of Machine dialog box

- Set desired parameters like how coolant NC code will be generated and how tool diameter & length offsets are generated in the machine program.
- You can create program in multiple subroutines by selecting respective option from the **Output subroutines** drop-down.
- Set the other parameters as desired and click on the **OK** button to apply the parameters.

Setup Options

Click on the **Setup** tab to define axis limits for the machine. The options will be displayed as shown in Figure-17. The options of this dialog box are discussed next.

Figure-17. Setup tab of Machine dialog box

- Select the None option from the Indexing drop-down to set the machine as 3-axis machine. Select the **4 Axis** option from the **Indexing** drop-down to set the machine as 4-axis. Select the **5 Axis** option from the drop-down to create 5-axis machine.
- Set desired rotary axis and tilt axis angular limits in respective sections of the dialog box.
- Select the **Update indexing angles for setups** check box to update the angular limits in NC program setup as well.
- Set the other parameters in this tab as discussed earlier.

You can set the other parameters of the dialog box as discussed earlier.

SETTING COORDINATE SYSTEM

The coordinate system is an important part of CNC machines programming. All the coordinates for machine programming are referenced to the coordinate system selected. The procedure to set coordinate system is given next.

- Right-click on the **Coordinate System** option from the **SOLIDWORKS CAM Feature Tree** and select the **Edit Definition** option from the shortcut menu; refer to Figure-18. The **Fixture Coordinate System PropertyManager** will be displayed; refer to Figure-19. Also, you will be asked to select a point to define the coordinate system.

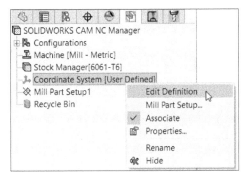

Figure-18. Edit Definition for Coordinate System

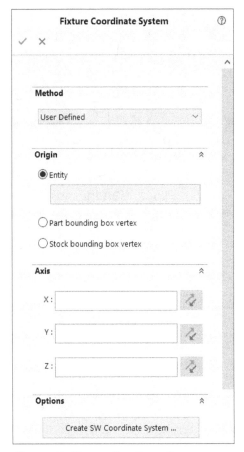

Figure-19. Fixture Coordinate System PropertyManager

- Select a point on the model that you want to be coordinate system. To align the coordinate system at desired orientation, click in the respective direction selection box of the **Axis** rollout and select the direction reference (face/plane/edge).
- If you want to use any vertex of the part bounding box then select the **Part bounding box vertex** radio button. Vertices will be displayed; refer to Figure-20.

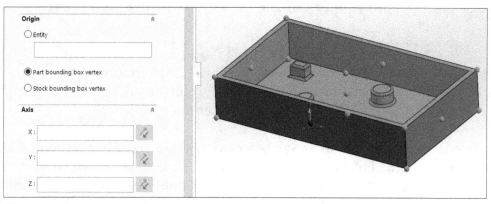

Figure-20. Bounding box vertices

- Select desired vertex to place coordinate system.
- If you want to use a vertex of bounding box of stock as coordinate system reference then select **Stock bounding box vertex** radio button.
- After selecting the point and setting the direction references, click on the **OK** button from the **PropertyManager** to create the coordinate system.

DEFINING OR EDITING STOCK

Stock is the pile of material to be removed after machining to produce the part. The procedure to define/edit the stock is given next.

- Click on the **Stock Manager** tool from the **SOLIDWORKS CAM CommandManager** in the **Ribbon**. The **Stock Manager PropertyManager** will be displayed; refer to Figure-21.

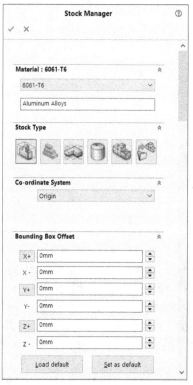

Figure-21. Stock Manager PropertyManager

- Select desired material from the drop-down.
- There are various options to set the stock type in the **Stock Type** rollout; **Bounding Box, Pre-defined Bounding Box, Cylindrical, Extruded Sketch, STL File**, and

Part File. Select the **Bounding Box** button if you want to create a stock bounding around the whole workpiece. You will be able to specify offset value for various faces when using this option.

- Select the **Pre-defined Bounding Box** button from the dialog box if you want to specify size of stock as well as offset values for creating stock.
- Select the **Extruded Sketch** button if you want to create stock by extruding the selected sketch; refer to Figure-22.
- Select the **Cylindrical** button to create a cylindrical stock around the model.
- Select the **STL File** button and click on the **Browse** button in the **STL File** rollout. The **Open** dialog box will be displayed; refer to Figure-23. Select desired file and click on the **Open** button.

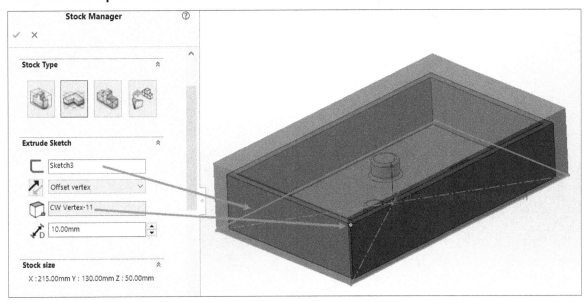

Figure-22. Extruded sketch stock

Figure-23. Open dialog box

- Click on the **Part File** button from the rollout if you want to use a SolidWorks file as stock. After selecting the button, click on the **Browse** button. The **Open** dialog box will be displayed. Select desired file and click on the **Open** button from the dialog box. The part will be placed at the origin.

*   After creating the stock and specifying desired parameters, click on the **OK** button. The stock will be created.

SETTING MILLING OPERATION PARAMETERS

The milling operations require a few parameters to be set before performing cutting operations. The procedure to set the parameters are given next.

*   Click on the **Mill Setup** tool from the **Setup** drop-down in the **SOLIDWORKS CAM TBM CommandManager**. The **Mill Setup PropertyManager** will be displayed; refer to Figure-24.

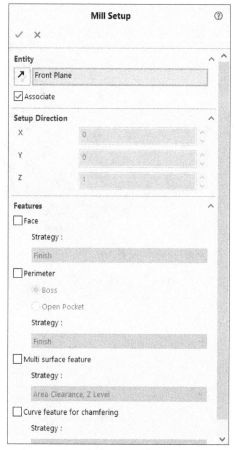

Figure-24. Mill Setup PropertyManager

*   Click in the selection box of **Entity** rollout and select the face/plane perpendicular to which the tool will be aligned while cutting. In simple words, select desired face/plane to define cutting direction.
*   Select the **Associate** check box if you want the setup to be modified with the model if any changes are made to it.
*   Select desired feature from the **Features** rollout. Like, select the **Face** check box if you want SolidWorks to automatically identify the face features for machining. Select the **Perimeter** check box if you want boss/open pocket features to be identified. Select the **Multi surface feature** check box if you want to identify multi surface features.
*   Click on the **OK** button from the **PropertyManager** to apply the settings.

EXTRACTING MACHINABLE FEATURES

The **Extract Machinable Features** tool is used to extract all the machinable features based on the cutting strategies in the database. The procedure to use this tool is given next.

* Right-click on **SOLIDWORKS CAM NC Manager** option from the **SOLIDWORKS CAM Feature Tree** and select the **Extract Machinable Features** option from the shortcut menu; refer to Figure-25 or click on the **Extract Machinable Features** tool from the **SOLIDWORKS CAM CommandManager** in the **Ribbon**. The identified features will be displayed; refer to Figure-26.

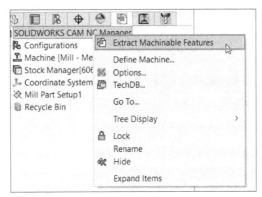

Figure-25. Extract Machinable Features option

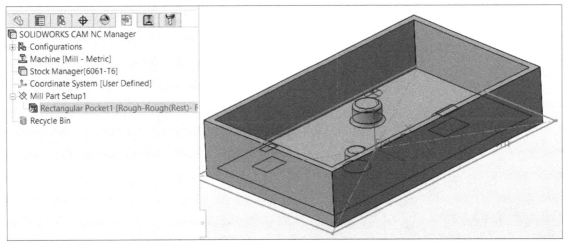

Figure-26. Automatically identified feature

GENERATING OPERATION PLAN

The **Generate Operation Plan** option is used to create cutting strategies based on identified features. To generate cutting strategies, right-click on desired machining feature and click on the **Generate Operation Plan** option from the shortcut menu or you can click on the **Generate Operation Plan** tool from the **Ribbon**. The material cutting strategies will be created; refer to Figure-27.

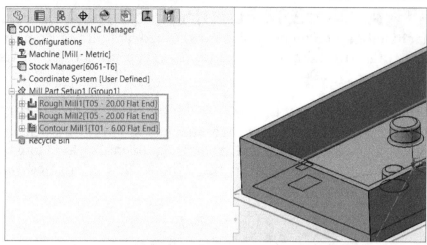

Figure-27. Material cutting strategies created

GENERATING TOOL PATHS

The **Generate Toolpath** option is used to create tool paths based on earlier generated operation plan. Select the operation plans from the **SOLIDWORKS CAM Operation Tree** and right-click on it. A shortcut menu will be displayed; refer to Figure-28. Select the **Generate Toolpath** option from the shortcut menu. The tool paths will be generated; refer to Figure-29.

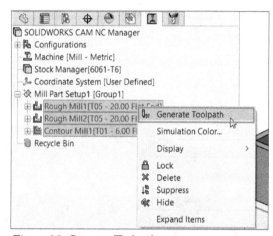

Figure-28. Generate Toolpath option

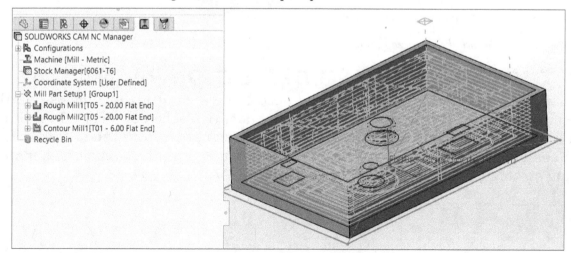

Figure-29. Toolpaths generated

SIMULATING TOOL PATHS

To simulate the toolpaths, select desired toolpath from the **SOLIDWORKS CAM Operation Tree** and right-click on it. A shortcut menu will be displayed. Select the **Simulate Toolpath** option from the shortcut menu; refer to Figure-30. The **Simulate Toolpath PropertyManager** will be displayed; refer to Figure-31.

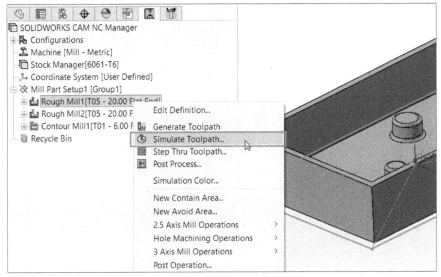

Figure-30. Simulate Toolpath option

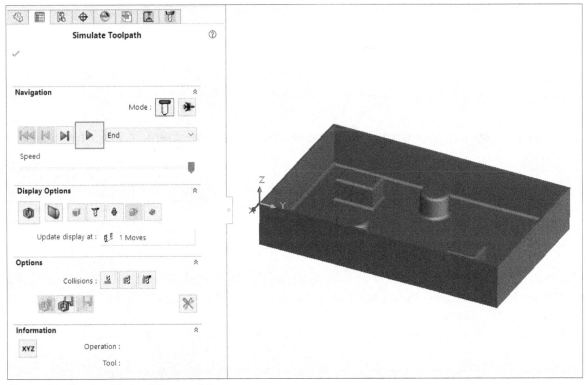

Figure-31. Simulate Toolpath PropertyManager

• Click on the **Play** button to simulate the toolpath. Set desired options in the **PropertyManager** and click on the **OK** button to exit the tool.

VISUALIZING TOOLPATHS

The **Step Thru Toolpath** option is used to check toolpaths at different instances of time. Select desired toolpath from the **SOLIDWORKS CAM Operation Tree** and right-click on the toolpath. A shortcut menu will be displayed; refer to Figure-30. Select the **Step Thru Toolpath** option from the shortcut menu. The **Step Through Toolpath PropertyManager** will be displayed; refer to Figure-32.

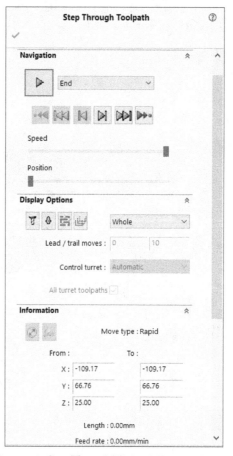

Figure-32. Step Through Toolpath PropertyManager

Click on the **Play** button from the **PropertyManager** to check each of the toolpath simulation.

SAVING TOOLPATHS

The **Save CL File** tool is used to save the toolpaths in CL file format. The procedure is given next.

- Click on the **Save CL File** tool from the **SOLIDWORKS CAM** cascading menu of **Tools** menu or from **SOLIDWORKS CAM CommandManager** in the **Ribbon**. The **Save As** dialog box will be displayed.
- Specify desired name and location. Click on the **Save** button to save the file.

POST PROCESSING TOOLPATHS

Post processing is the process of generating G codes from the toolpaths. The procedure to do so is given next.

- Click on the **Post Process** tool from the **SOLIDWORKS CAM** cascading menu of the **Tools** menu or from the **Ribbon**. The **Post Output File** dialog box will be displayed; refer to Figure-33.

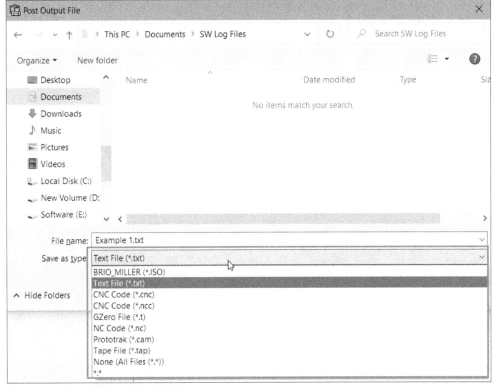

Figure-33. Post Output File dialog box

- Specify desired name and file type, and click on the **Save** button. The **Post Process PropertyManager** will be displayed; refer to Figure-34.
- Select the **Open G-Code file in** check box from **Options** rollout if you want to edit the G-codes after creating them.
- Select the **Centerline** check box to display movement of centerline of cutting tool in simulation when posting the codes.
- Click on the **Play** button from the **PropertyManager** to start creating G-codes. The codes will be displayed in the **NC Codes** box.

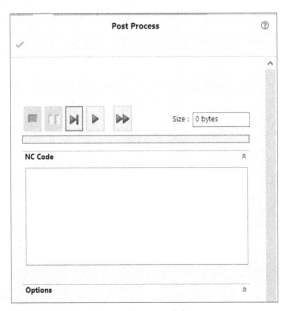

Figure-34. Post Process PropertyManager

- Click on the **OK** button from the **PropertyManager**. The **SOLIDWORKS CAM NC Editor** will be displayed if you have selected the **Open G-Code file in** check box; refer to Figure-35.

•   Modify the file as required and save it. Close the application

Figure-35. SOLIDWORKS CAM NC Editor

MANUALLY CREATING MILL OPERATIONS

In previous sections, you have learned to automatically create milling operations. Now, we will discuss the procedure of creating different milling operations manually. We will work with Rough Mill Operation and you can apply the same procedure to other tools.

•   Right-click on **Mill Part Setup1** option from the **SOLIDWORKS CAM Operation Tree** and select **Rough Mill** option from **2.5 Axis Mill Operations** cascading menu in shortcut menu; refer to Figure-36. The **New Operation : Rough Mill PropertyManager** will be displayed; refer to Figure-37.

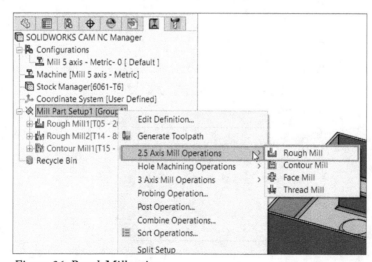

Figure-36. Rough Mill option

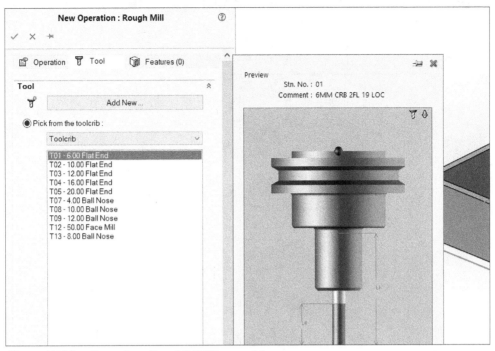

Figure-37. New Operation : Rough Mill PropertyManager

- Select desired operation type from the drop-down at the top in the **Operation** tab of the **PropertyManager**. Select the **Rough Mill** option to perform rough milling operation for large stock. Select the **Contour Mill** option to remove material around the edges of boss/pocket features. Select the **Face Mill** option to remove material from the top face of a part. Select the **Thread Mill** option to machine threads on boss or hole features. We are using the **Rough Mill** option in our case; refer to Figure-38.

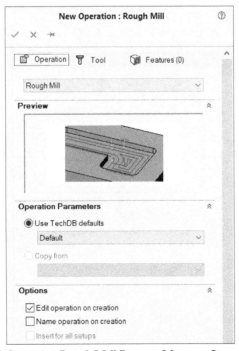

Figure-38. New Operation Rough Mill PropertyManager Operation tab

- Click on the **Tool** tab in the **PropertyManager**. The options will be displayed as shown in Figure-37.
- Select desired tool from the list.

- Click on the **Features** tab in the **PropertyManager**. The options will be displayed as shown in Figure-39.

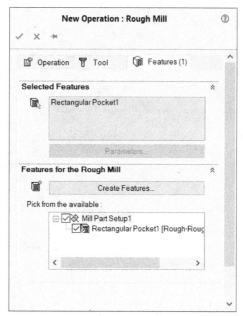

Figure-39. Features tab in New Operation PropertyManager

- Select desired feature from the **Pick from the available** box in the **Features for the Rough Mill** rollout if you have earlier applied **Extract Machinable Features** tool. If you have not used the tool earlier then you can create features now also.
- Click on the **Create Features** button from the **PropertyManager** and then click on the **2.5 Axis Feature** option. The **2.5 Axis Feature: Select Entities PropertyManager** will be displayed; refer to Figure-40.

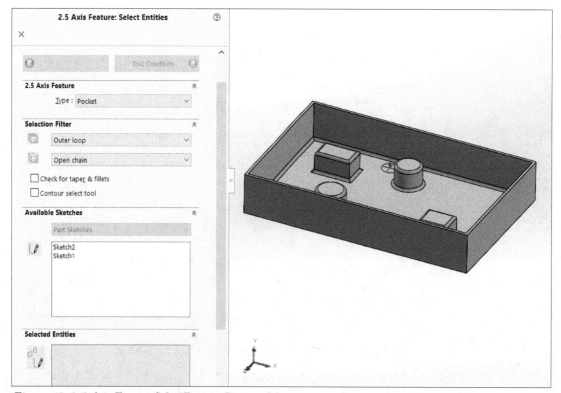

Figure-40. 2.5 Axis Feature Select Entities PropertyManager

- Select desired feature type from the **Type** drop-down of **2.5 Axis Feature** rollout which is **Pocket** in our case. Set the selection filters as required from the **Selection Filter** rollout.
- Click on the face(s) to be used to create the feature; refer to Figure-41.

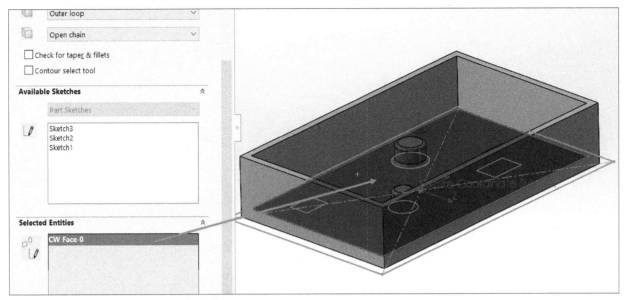

Figure-41. Face selected for creating mill features

- Click on the **End Condition** button at the top in the **PropertyManager**. The options will be displayed as shown in Figure-42.

Figure-42. End condition options

- Set desired parameters and click on the **OK** button. The feature will be created.

- After creating the feature and selecting it, click on the **OK** button from the **New Operation PropertyManager**. The cutting operation will be created and the **Operation Parameters** dialog box will be displayed; refer to Figure-43.
- Set the parameters as needed and click on the **OK** button to create the operation.
- You can now generate the toolpath as discussed earlier.

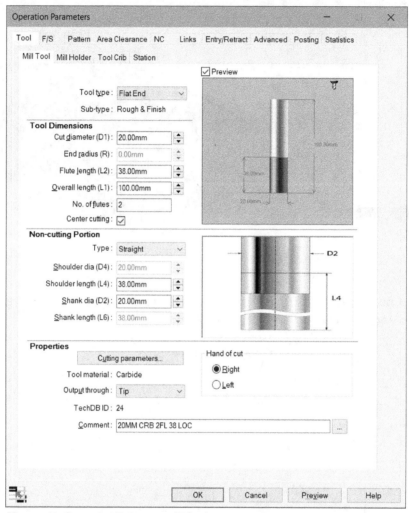

Figure-43. Operation Parameters dialog box

Till this point, we have learned about the basics of SolidWorks CAM. In the next chapters, we will discuss each of the options in detail.

SELF ASSESSMENT

Q1. Discuss the difference between roughing process and finishing process.

Q2. Which of the following is an unconventional machine?

a. Electric Discharge Machine (EDM)   b. Vertical Milling Machine
c. Shaper Machine   d. Planar Machine

Q3. Discuss the applications of CAM in manufacturing.

Q4. Draw the flow chart of basic approach for CAM program generation.

Q5. Post processor is used to translate the codes generated by CAM program to machine readable codes based on machine controller. (T/F)

Q6. Define the term "stock" for machining.

Q7. Which of the following buttons is used to import model of stock?

a. Bounding Box   b. Pre-defined Bounding Box
c. STL File   d. Extruded sketch

Q.8. Which of the following tools is used to specify cutting strategy for machining features?

a. Generate Operation Plan   b. Extract Machinable Features
b. Mill Setup   d. Generate Toolpath

Q9. Which of the following tools is used to check the simulation of material removal from workpiece?

a. Step Thru Toolpath   b. Simulate Toolpath
c. Post Process   d. Save CL File

Chapter 2

Setting Up Workpiece and Machine

Topics Covered

The major topics covered in this chapter are:

- *Introduction.*
- *Opening/Importing Part.*
- *CNC Machine Structure.*
- *Cutting Tools used in CNC Milling and Turning.*
- *Defining Stock for Part.*

INTRODUCTION

The first step in using CAM software is importing/creating solid model which is digital copy of the part you want to manufacture. After importing, you need to create setup for machining the part. The process to import model and setup machine is discussed next.

OPENING/IMPORTING PART

The **Open** tool in SolidWorks is used to open/import model for machining. The procedure to use this tool is given next.

- Click on the **Open** tool from the **Quick Access Toolbar**, press **CTRL+O**, or click on the **Open** tool from the **File** menu. The **Open** dialog box will be displayed; refer to Figure-1.

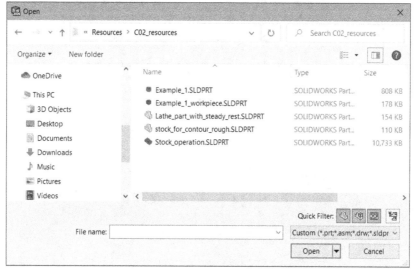

Figure-1. Open dialog box

- Select desired option from the **Files of type** drop-down; refer to Figure-2. The files of selected format will be displayed in the **Open** dialog box.

Figure-2. Files of Type drop-down

- Select desired file from the dialog box and click on the **Open** button. The model will be displayed in the application window; refer to Figure-3.

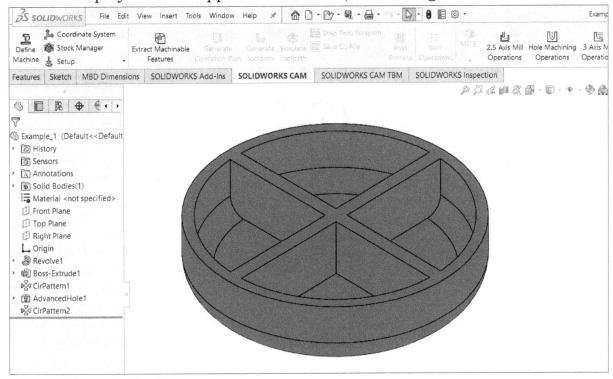

Figure-3. File opened in SolidWorks

CNC MACHINE STRUCTURE

As discussed in the starting of book, CNC machines are the material cutting machines which use numeric codes to perform action. These numeric codes are understood by controller installed on the machine which gives command to various motors in the machine. But, this is not our concern now. Now, we will discuss what are the components of machine that we should know before we create or change a machine in SolidWorks CAM. Figure-4 shows a 5 axis VMC with some of the components. Various major components milling and lathe machines are discussed next.

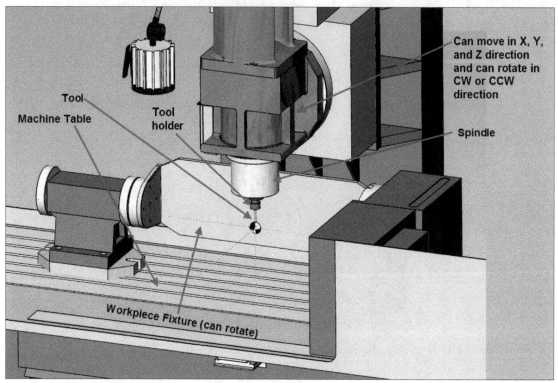

Figure-4. Five axis VMC

Tool Spindle

Machine tool spindles are rotating components that are used to hold and drive cutting tools or work pieces on lathes, milling machines and other machine tools. They use belt, gear, motorized, hydraulic or pneumatic drives and are available in a variety of configurations. Various specifications for selecting tool spindle are given next.

- Select spindle as per the required spindle Speed.
- Make sure spindle orientation correct as per your application.
- Consider the gage length of the tool. Doubling the gage length of a tool can increase the deflection at the end by a factor of 8. A way to compensate for this would be to go from a 40 taper to a 50 taper spindle and tool holder.
- Choose a spindle that can transmit the required amount of power/torque.
- When boring, select a spindle that has a nose bearing ID larger than the bore being machined
- Select the nose bearing arrangement suited for the application

Tool Changer or Turret

Tool changer, tool indexer or tool turret is used to automatically change the tool in spindle; refer to Figure-5. In some CNC milling and Lathe machines, you can load more than one tool at a time and then use the NC codes to use them in different toolpaths. While purchasing the machine, you should keep a note of time taken by machine to automatically change the tool and direction in which turret can rotate.

Figure-5. Tool Spindle with indexer

Translational and Rotational Limits

The translational and rotational limits are important aspect of CNC machines. You can not make a job which required machining length or angle more than the translational or rotational limits of machine. You can find this information in catalog of machine.

Tail Stock

Tail stock is generally found in lathe machines but can also be seen with rotary table of a milling machine. Tail stock is used to support long workpiece at its end; refer to Figure-6.

Figure-6. Tail Stock in lathe

DEFINING MACHINE

The **Define Machine** tool is used to setup parameters related to machine to be used for CAM data generation. The procedure to use this tool is given next.

• Click on the **Define Machine** tool from the **SOLIDWORKS CAM CommandManager** in the **Ribbon**. The **Machine** dialog box will be displayed; refer to Figure-7.

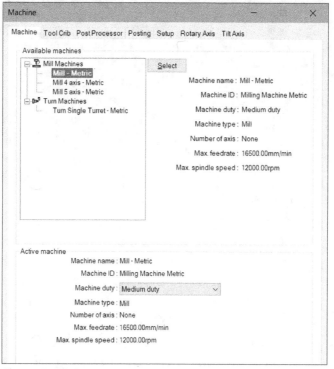

Figure-7. Machine dialog box

Machine tab

The **Machine** tab is selected by default. The options in this tab are used to select machine for model and set related parameters.

• Select desired option from the **Available machines** area of the dialog box and click on the **Select** button. The machine will be set as active machine for current CAM project. There are mainly two types of machines; mill machine and turn machine.
• Select desired option from the **Machine duty** drop-down to define how aggressively the machine will be running during operation. Selecting the **Heavy duty** option from the drop-down will give aggressive speeds and feeds suited for a large and rigid machine. Select the **Light duty** option from the drop-down if you want to machine small parts with hard material or the machine is not rigid enough for high feed. Based on the option selected in this drop-down, the values of speed and feedrate are selected from the database.

Tool Crib tab

The options in the **Tool Crib** tab are used to setup cutting tools to be used for machining; refer to Figure-8. The options in this tab are discussed next.

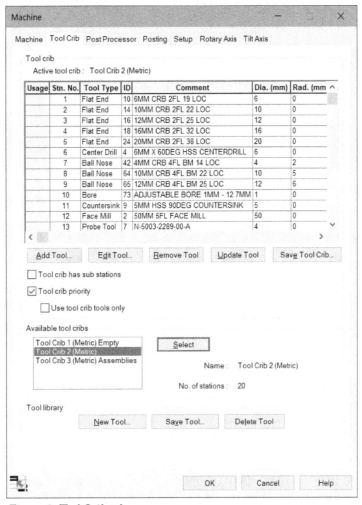

Figure-8. Tool Crib tab

- Select the predefined tool crib from the **Available tool cribs** area and click on the **Select** button if you want to use predefined cribs.
- The number of tools that can be added to the crib are displayed in **No. of stations** field of **Available tool cribs** area. Note that you need to check with your machine manufacturer data also to find out how many tools can be added to the crib. The tool crib represents turret in the language of Machinists.

Adding Tool to Crib

- Click on the **Add Tool** button from the **Tool crib** area. The **Tool Select Filter** dialog box will be displayed; refer to Figure-9.

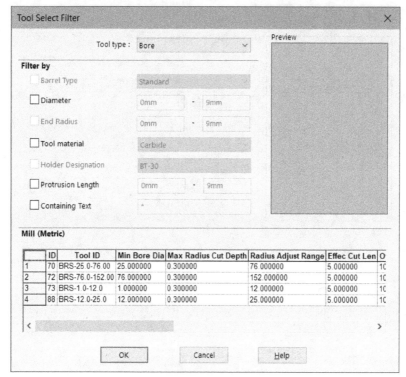

Figure-9. Tool Select Filter dialog box

- Select desired tool type from the **Tool type** drop-down. The list of tools of selected type will be displayed in the table.
- Select the **Diameter** check box to specify range diameter within which you want to select the tool and specify the parameters in the edit boxes.
- You can set the other filter parameters like tool material, end radius, holder designation, protrusion length, and so on in the same way.
- After setting filters, select desired tool from the list and click on the **OK** button. The tool will be added in the **Tool crib** area of the dialog box.

Editing Tool of Tool Crib

- Select the tool from the table which you want to be edited. Preview of tool will be displayed; refer to Figure-10.

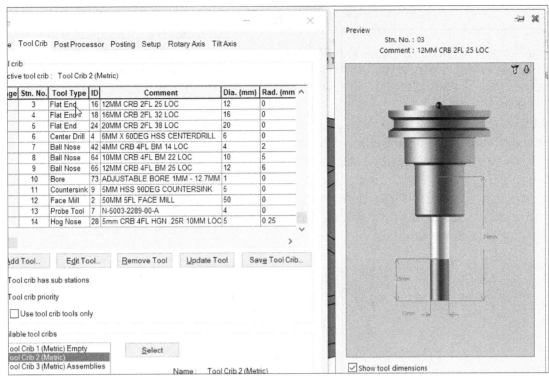

Figure-10. Preview of tool

- Click on the **Edit Tool** button from the dialog box. The **Edit Tool Parameters** dialog box will be displayed; refer to Figure-11.

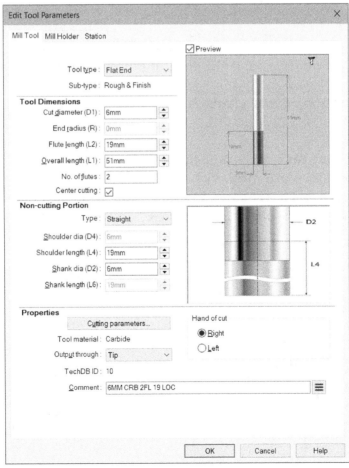

Figure-11. Edit Tool Parameters dialog box

Mill Tool tab

- Select desired tool type from the **Tool type** drop-down. The parameters in the dialog box will be displayed accordingly.
- Specify desired parameters in the **Tool Dimensions** area to define size and cutting parameters of tool like cut diameter, end radius, flute length, number of flutes, and so on.
- In **Non-cutting Portion** area, you need to specify parameters related to non-cutting portion of tool. Select desired option from the **Type** drop-down to define shape of non-cutting length. Select the **Straight** option to create straight cutting length. Select the **Tapered** option to create shoulder length in taper. Select the **Neck** option from the **Type** drop-down to define stepped non-cutting length. The parameters based on your selection will be displayed as shown in Figure-12. Set the parameters as desired.

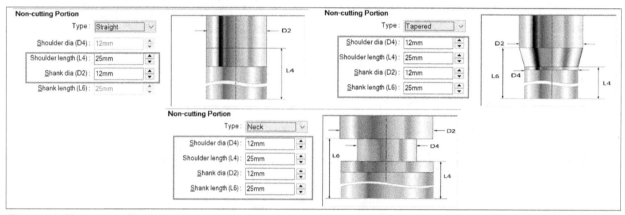

Figure-12. Non-cutting Portion options

- Select desired radio button from the **Hand of cut** area to define cutting orientation of tool. You can define the tool right-handed or left handed by selecting respective radio button.
- Click on the **Cutting parameters** button from the **Properties** area to define cutting parameters of tool. The **Cutting Parameters** dialog box will be displayed; refer to Figure-13.
- By default, the **Associate with stock material** check box is selected and hence the cutting parameters are applied based on stock material.
- Clear the **Associate with stock material** check box to define cutting parameters associated with tool. The parameters in the dialog box will become active.

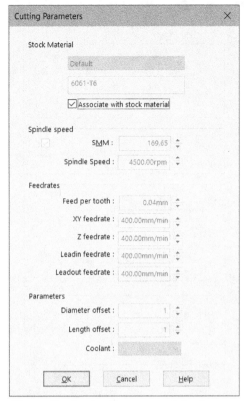

Figure-13. Cutting Parameters dialog box

- Select the **SMM** check box to define surface meter per minute speed of spindle. Clear this check box if you want to specify speed of spindle in RPM (rounds per minute). Spindle speed is the rotational speed of spindle at which cutting tool will rotate while cutting in case of milling machine and rotational speed at which workpiece will rotate in lathe-turning machine.

- Specify the distance travelled by tool per tooth rotation in the **Feed per tooth** edit boxes.

- Specify desired values in the **XY feedrate** and **Z feedrate** edit boxes to define feedrate in XY plane and Z direction, respectively.

- Specify desired values in **Leadin feedrate** and **Leadout feedrate** edit boxes to define the rate at which tool will move in and move out of the workpiece while cutting, respectively. Note that specified leadout feedrate also applies to retract parameter.

- After setting desired parameters, click on the **OK** button from the dialog box. The **Edit Tool Parameters** dialog box will be displayed again.

- Select desired option from the **Output through** drop-down. Select the **Tip** option if you want the codes to be generated based on movement of tool tip. Select the **Center** option from the drop-down if you want to use the center of tool for NC code generation.

- In the **Comment** edit box, specify desired text to define name of tool. Click on the **Additional Details** button to specify more details about tool like manufacturer, ID, and so on. On clicking this button, the **Additional Details (Tool/Holder)** dialog box will be displayed; refer to Figure-14. Specify desired parameters and click on the **OK** button.

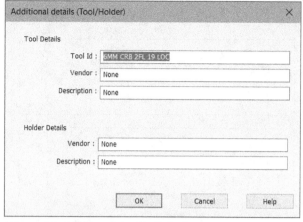

Figure-14. Additional details dialog box

Mill Holder tab

- Click on the **Mill Holder** tab from the dialog box. The options will be displayed as shown in Figure-15.

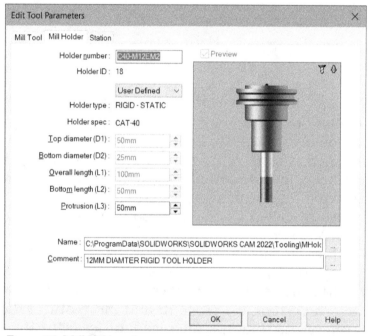

Figure-15. Mill Holder tab

- Specify desired name/number in the **Holder number** edit box.
- Select desired option from the drop-down to define shape of holder. Select the **Basic** option from the drop-down to create basic shape tool holder. Select the **User Defined** option from the drop-down to use a user-defined shape.
- Set desired parameters for defining shape of holder like top diameter, bottom diameter, overall length, and so on.
- If the **User Defined** option is selected in the drop-down then click on the **Browse** button for **Name** field in the dialog box. The **Open** dialog box will be displayed for opening mill holder file (*.mh). Select desired file and click on the **Open** button. New mill tool holder will be displayed in the dialog box with respective parameters.
- Specify desired description in the **Comment** edit box or click on the **Browse** button for **Comment** edit box. The **Additional details (Tool/Holder)** dialog box will be displayed; refer to Figure-16.

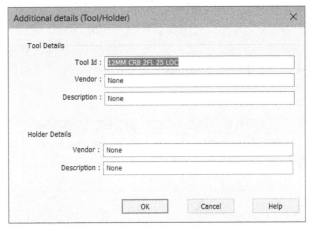

Figure-16. Additional details dialog box

- Specify desired values in the edit boxes and click on the **OK** button. The parameters will be updated in the **Comment** edit box.

Station Tab

- Click on the **Station** tab to define tool station parameters. The options will be displayed as shown in Figure-17.

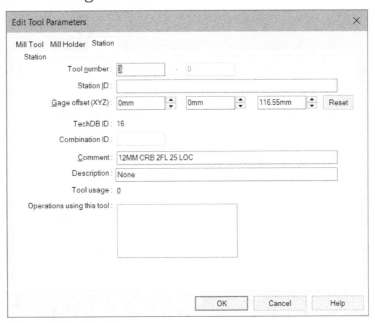

Figure-17. Station tab

- Specify desired value in **Tool number** edit box. The tool will be placed at specified number in the turret (tool crib).
- Click in the **Station ID** edit box and specify desired value. Note that a machine can have multiple stations and substations for holding the cutting tools.
- Specify the other parameters as desired and click on the **OK** button. The tool will be updated in the **Tool Crib** tab of **Machine** dialog box.

Removing Tool from Crib

The **Remove Tool** button is used to remove a cutting tool earlier added in the crib. The procedure to do so is given next.

- Select the cutting tool to be deleted from the table in **Tool crib** area and click on the **Remove Tool** button. The selected cutting tool will be removed from the table.

The **Update Tool** button is used to update the tool data of selected tool as per the technical database.

Saving Tool Crib to Database

The **Save Tool Crib** button is used to save the current crib for later use. On clicking this button, the **Save to Database** dialog box will be displayed; refer to Figure-18.

Figure-18. Save to Database dialog box

- Specify desired name for tool crib in the **Name** edit box.
- Select the **Add as new tool** check box to add tools of crib to the database as new tools.
- Select the **Update tool** check box to update changes in the cutting tools to database.
- Select the **Add new tools to database** check box to add all new tools of crib to the database.
- After setting parameters, click on the **Save** button. The new crib will be added in the list of available tool cribs.
- The options in the **Tool library** area of the dialog box are used to create, save, and delete cutting tools.

Post Processor Tab

The options in the post processor tab are used to set post processor parameters like controller make, name of machine, and so on; refer to Figure-19. The options in this tab are discussed next.

- Select desired option from the list of post processors in the **Available** area of the dialog box. If desired post processor is not available in the list then click on the **Browse** button. The **Open** dialog box will be displayed; refer to Figure-20.

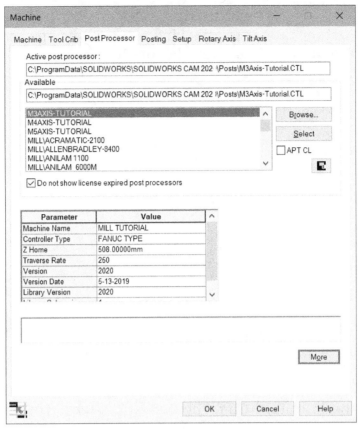

Figure-19. Post Processor tab

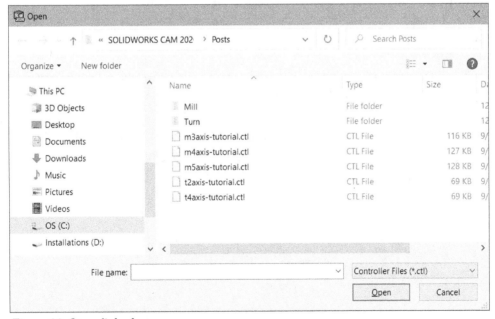

Figure-20. Open dialog box

- Select desired option from the dialog box and click on the **Open** button.
- After selecting desired post processor from the list, click on the **Select** button from the **Available** area. The selected post processor will be set as active post processor for the project.
- Select the **APT CL** check box if you want to use external post processor specific to the machine. After selecting this check box, click on the Browse button to use controller file with extension *.APTctl.

Posting Tab

The options in the **Posting** tab are used to define offset and coolant parameters; refer to Figure-21. The options in this tab are discussed next.

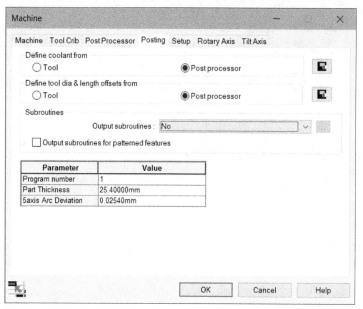

Figure-21. Posting tab

- Select the **Tool** radio button from the **Define coolant from** area to set coolant parameters as per cutting tool settings. Select the **Postprocessor** radio button from the **Define coolant from** area to set the coolant parameters as per the post processor settings.
- Similarly, you can set the tool offset in **Define tool dia & length offsets from** area of the dialog box.
- Set desired parameters in the table at the bottom in the dialog box by double-clicking in the field.

Setup Tab

The options in the **Setup** tab are used to define coordinate system, tilt axis, rotary axis, and so on; refer to Figure-22. The options in this tab are discussed next.

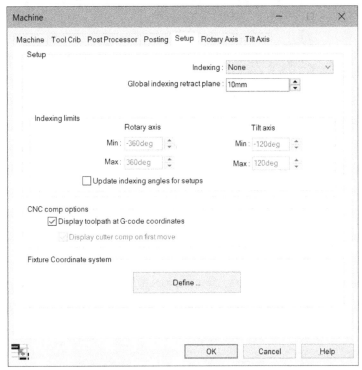

Figure-22. Setup tab

- Select desired option from the **Indexing** drop-down. Select the **4 Axis** option if the workpiece can rotate. Select the **5 Axis** option if the tool can also tilt along with rotation of workpiece. Based on selected option, the parameters in the Indexing limits area will become active.
- Specify desired value in the **Global indexing retract plane** edit/spinner to define the distance in Z direction upto which the tool will move before next cutting operation.
- Specify desired values in the **Rotary axis** and **Tilt axis** spinners to define limits of workpiece rotation and tool tilting.
- Select the **Update indexing angles for setups** check box if maximum and minimum range of machine changes.
- Select the **Display toolpath at G-code coordinates** check box if you want the toolpath displayed on coordinates generated by CAM data. De-select the check box if you want to use general machine toolpath for simulation.
- Click on the **Define** button from the **Fixture Coordinate system** area of the dialog box. The **Fixture Coordinate System PropertyManager** will be displayed; refer to Figure-23.

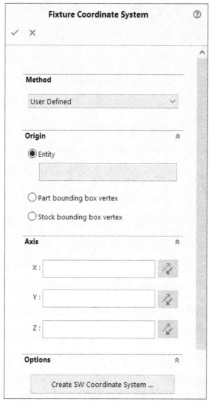

Figure-23. Fixture Coordinate System PropertyManager

- Select desired option from the **Method** drop-down. You can use either SolidWorks coordinate system by using the **SOLIDWORKS Coordinate System** option or you can create SolidWorks CAM coordinate system option by selecting the **User Defined** option.

User Defined Coordinate System

- Select the **Entity** radio button if you want to select vertex on a model to define location of coordinate system. Select the **Part bounding box vertex** radio button if you want to select a vertex of part bounding box for placing coordinate system; refer to Figure-24. Select the **Stock bounding box vertex** radio button to use vertex of stock.

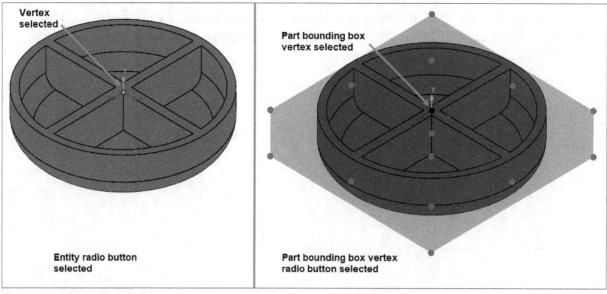

Figure-24. Selecting vertex for coordinate system

- Click in the **X** selection box of **Axis** rollout and select an edge/face for defining direction of X axis of coordinate system.
- Similarly, you can use Y and Z selection boxes to define direction of Y and Z axes of the coordinate system.

SOLIDWORKS Coordinate System

- Select the **SOLIDWORKS Coordinate System** option from the **Method** drop-down in the **PropertyManager**. You will be asked to select a SolidWorks coordinate system.
- Select desired SolidWorks coordinate system from the **Available Coordinate Systems** selection box; refer to Figure-25 and click on the **OK** button from the **PropertyManager**.

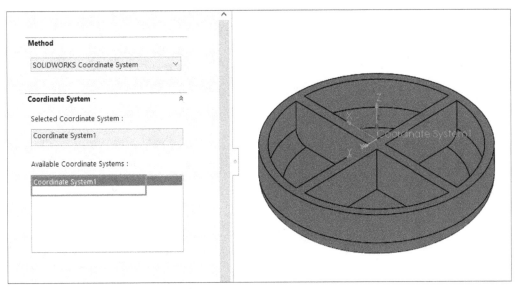

Figure-25. Coordinate system selected

Rotary Axis Tab

The options in the **Rotary Axis** tab are used to define the axis and rotation direction for rotary axis; refer to Figure-26. The options in this tab are discussed next.

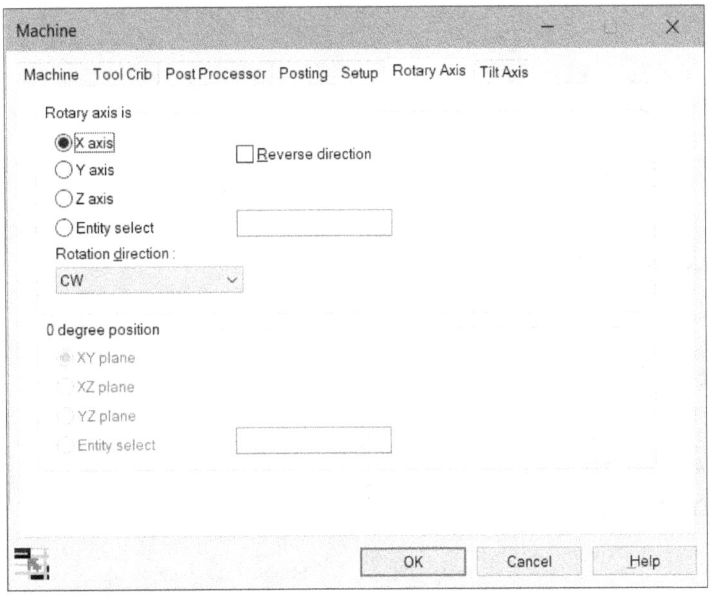

Figure-26. Rotary Axis tab

- Select desired radio button from the **Rotary axis is** area to define axis about which workpiece can rotate. If you have selected the **Entity select** radio button then you need to select an edge/axis/cylindrical face to define axis.
- Select desired option from the **Rotation direction** drop-down to define rotation about the selected axis. You can select clockwise (CW), counter-clockwise (CCW), or Both option.

Tilt Axis Tab

The options in the **Tilt Axis** tab are used to define parameters for cutting tool tilt when performing 4-axis or 5-axis milling; refer to Figure-27.

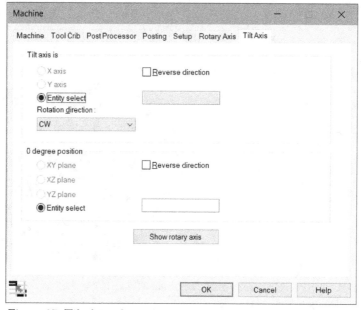

Figure-27. Tilt Axis tab

- Set desired parameters as discussed earlier and click on the **Show rotary axis** button. The preview of axes will be displayed; refer to Figure-28.

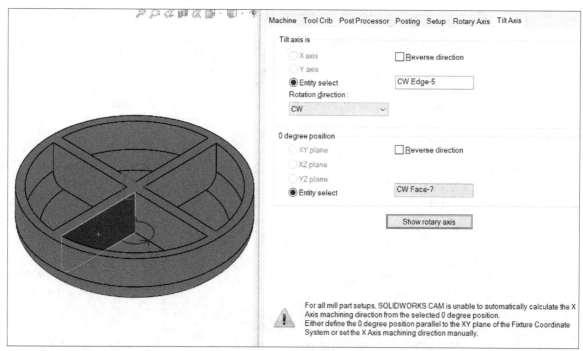

Figure-28. Preview of axes

• Click on the **OK** button from the dialog box to apply settings.

TOOLS USED IN CNC MILLING AND LATHE MACHINES

The tools used in CNC machines are made of cemented carbide, High Speed Steel, Tungsten Alloys, Ceramics, and many other hard materials. The shapes and sizes of tools used in Milling machines and Lathe machines are different from each other. These tools are discussed next.

Milling Tools

There are various type of milling tools for different applications. These tools are discussed next.

End Mill

End mills are used for producing precision shapes and holes on a Milling or Turning machine. The correct selection and use of end milling cutters is paramount with either machining centers or lathes. End mills are available in a variety of design styles and materials; refer to Figure-29.

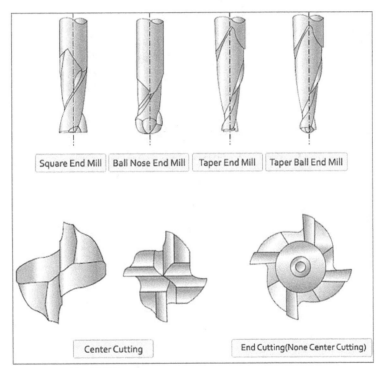

Figure-29. End Mill tool types

Titanium coated end mills are available for extended tool life requirements. The successful application of end milling depends on how well the tool is held (supported) by the tool holder. To achieve best results an end mill must be mounted concentric in a tool holder. The end mill can be selected for the following basic processes:

FACE MILLING - For small face areas, of relatively shallow depth of cut. The surface finish produced can be 'scratchy".
KEYWAY PRODUCTION - Normally two separate end mills are required to produce a quality keyway.

WOODRUFF KEYWAYS - Normally produced with a single cutter, in a straight plunge operation.

SPECIALTY CUTTING - Includes milling of tapered surfaces, "T" shaped slots & dovetail production.

FINISH PROFILING - To finish the inside/outside shape on a part with a parallel side wall.

CAVITY DIE WORK - Generally involves plunging and finish cutting of pockets in die steel. Cavity work requires the production of three dimensional shapes. A Ball type end mill is used for the finishing cutter with this application.

Roughing End Mills, also known as ripping cutters or hoggers, are designed to remove large amounts of metal quickly and more efficiently than standard end mills; refer to Figure-30. Coarse tooth roughing end mills remove large chips for heavy cuts, deep slotting and rapid stock removal on low to medium carbon steel and alloy steel prior to a finishing application. Fine tooth roughing end mills remove less material but the pressure is distributed over many more teeth, for longer tool life and a smoother finish on high temperature alloys and stainless steel.

Figure-30. Roughing End Mill

Bull Nose Mill

Bull nose mill look alike end mill but they have radius at the corners. Using this tool, you can cut round corners in the die or mold steels. Shape of bull nose mill tool is given in Figure-31.

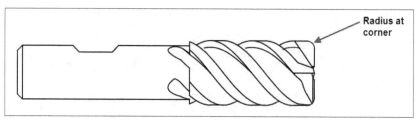

Figure-31. Bull Nose Mill cutter

Ball Nose Mill

Ball nose cutters or ball end mills have the end shape hemispherical; refer to Figure-29. They are ideal for machining 3-dimensional contoured shapes in machining centres, for example in moulds and dies. They are sometimes called ball mills in shop-floor slang. They are also used to add a radius between perpendicular faces to reduce stress concentrations.

Face Mill

The Face mill tool or face mill cutter is used to remove material from the face of workpiece and make it flat; refer to Figure-32.

Figure-32. Face milling tool

Radius Mill and Chamfer Mill

The Radius mill tool is used to apply round (fillet) at the edges of the part. The Chamfer mill tool is used to apply chamfer at the edges of the part. Figure-33 shown the radius mill tool and chamfer mill tool.

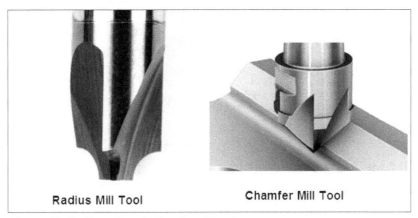

Radius Mill Tool   Chamfer Mill Tool

Figure-33. Radius mill and Chamfer mill tool

Slot Mill

The Slot mill tool is used to create slot or groove in the part metal. Figure-34 shows the shape of slot mill tool.

Figure-34. Slot mill tool

Taper Mill

In CNC machining, taper end mills are used in many industries for a large number of applications, such as walls with draft or clearance angle, tool and die work, mold work, even for reaming holes to make them conical. There are mainly two types of taper mills, Taper End Mill and Taper Ball Mill; refer to Figure-29.

Dove Mill

Dove mill or Dovetail cutters are designed for cutting dovetails in a wide variety of materials. Dovetail cutters can also be used for chamfering or milling angles on the bottom surface of a part. Dovetail cutters are available in a wide variety of diameters and in 45 degree or 60 degree angles; refer to Figure-35.

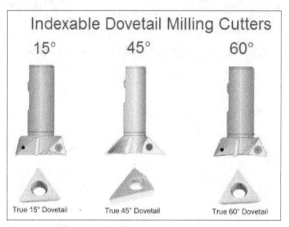

Figure-35. Dovetail milling cutters

Lollipop Mill

The Lollipop mill tool is used to cut round slot or undercuts in workpiece. Some tool suppliers use a name Undercut mill tool in place of Lollipop mill in their catalog. The shape of lollipop mill tool is given in Figure-36.

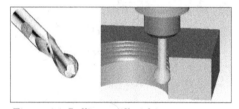

Figure-36. Lollipop mill tool

Engrave Mill

The Engrave mill tool is used to perform engraving on the surface of workpiece. Engraving has always been an art and it is also true for CNC machinist. You can find various shapes of engraving tool that are single flute or multi-flute; refer to Figure-37. You can use ball mill/end mill for engraving or you can use specialized engrave mill tool for engraving. This all depends on your requirement. If you want to perform engraving on softer materials or plastics then it is better to use ball end mill but if you want an artistic shade on the surface then use the respective engrave mill tool. Keep a note of maximum depth and spindle speed mentioned by your engrave mill tool supplier.

Multi flute Engrave End mill   Diamond Shaped Engrave Mill

Figure-37. Engrave mill tools

Thread Mill

The Thread mill tool is used to generate internal or external threads in the workpiece. The most common question here is if we have Taps to create thread then why is there need of Thread mill tool. The answer is less machining time on CNC, tool cost saving, more parts per tool, and better thread finish. Now, you will ask why to use tapping. The answer is low machine cost. Figure-38 shows thread mill tools.

Figure-38. Thread Mill

Barrel Mill

Barrel Mill tool is the tool recently being highly used in machining turbine/impeller blades and other 5-axis milling operations. Barrel Mill has conical shape with radius at its end; refer to Figure-39. Note that earlier Ball mill tools were used for irregular surface contouring but Barrel Mill tools give much better surface finish so they are highly in demand for 5-axis milling now a days.

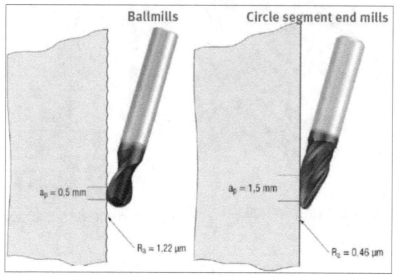

Figure-39. Barrel Mill versus Ball Mill Tool

Drill Bit

Drill bit is used to make a hole in the workpiece. The hole shape depends on the shape of drill bit. Drill bits for various purposes are shown in Figure-40. Note that drill is the machine or holder in which drill bit is installed to make cylindrical holes. There are mainly four categories of drill bit; Twist drill bit, Step drill bit, Unibit (or conical bit), and Hole Saw bit (Refer to Figure-41). Twist drill bits are used for drilling holes in wood, metal, plastic and other materials. For soft materials the point angle is 90 degree, for hard materials the point angle is 150 degree and general purpose twist drill bits have angle of 150 degree at end point. The Step drill bits are used to make counter bore or countersunk holes. The Unibits are generally used for drilling holes in sheetmetal but they can also be used for drilling plastic, plywood, aluminium and thin steel sheets. One unibit can give holes of different sizes. The Hole saw bit is used to cut a large hole from the workpiece. They remove material only from the edge of the hole, cutting out an intact disc of material, unlike many drills which remove all material in the interior of the hole. They can be used to make large holes in wood, sheet metal and other materials.

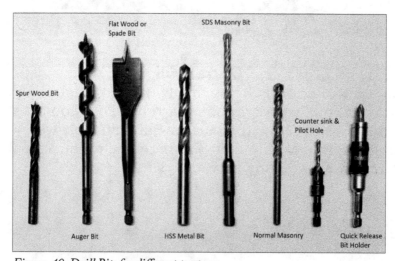

Figure-40. Drill Bits for different purposes

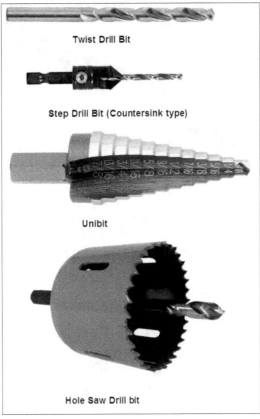

Twist Drill Bit

Step Drill Bit (Countersink type)

Unibit

Hole Saw Drill bit

Figure-41. Types of drill bits

Reamer

Reamer is a tool similar to drill bit but its purpose is to finish the hole or increase the size of hole precisely. Figure-42 shows the shape of a reamer.

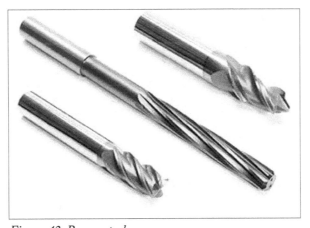

Figure-42. Reamer tool

Bore Bar

Bore Bar or Boring Bar is used to increase the size of hole; refer to Figure-43. One common question is why to use bore bar if we can perform reaming or why to perform reaming when we have bore bar. The answer is accuracy. A reamer does not give tight tolerance in location but gives good finish in hole diameter. A bore bar gives tight tolerance in location but takes more time to machine hole as compared to reamer. The decision to choose the process is on machinist. If you need a highly accurate hole then perform drilling, then boring and then reaming to get best result.

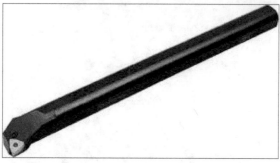

Figure-43. Boring Bar

Lathe Tools or Turning Tools

The tools used in CNC lathe machines use a different nomenclature. In CNC lathe machines, we use insert for cutting material. The Insert Holder and Inserts have a special nomenclature scheme to define their shapes. First we will discuss the nomenclature of Insert holder and then we will discuss the nomenclature of Inserts.

Insert Holders

Turning holder names follow an ISO nomenclature standard. If you are working on a CNC shop floor with lathes, knowing the ISO nomenclature is a must. The name looks complicated, but is actually very easy to interpret; refer to Figure-44.

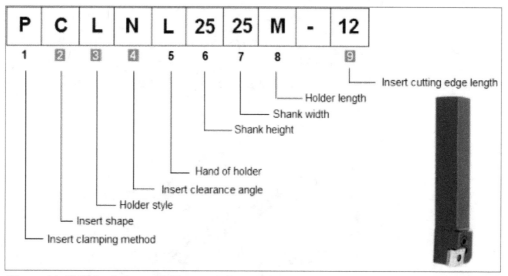

Figure-44. CNC Lathe Insert Holder nomenclature

When selecting a holder for an application, you mainly have to concentrate on the numbers marked in red in the above nomenclature. The others are decided automatically (e.g., the shank width and height are decided by the machine), or require less effort. In Figure-45, the rows with the question mark indicate the parameters that require the decision by machinist based on job.

	Parameter		How is this decided ?
1	Insert clamping method		Select based on cutting forces. Top clamping is the most sturdy, screw clamping the least.
2	Insert shape	?	Decided by the contour that you want to turn.
3	Holder style	?	Decided by the contour that you want to turn.
4	Insert clearance angle	?	Positive / Negative, based on application.
5	Hand of holder		Decided based on whether you want to cut towards the chuck or away from the chuck, and on turret position - turret front / rear
6	Shank height		Decided by holder size.
7	Shank width		Decided by machine.
8	Holder length		Decided by machine.
9	Insert cutting edge length	?	Decide based on depth of cut you want to use.

Figure-45. CNC Lathe Insert Holder nomenclature parameters

Figure-46 and Figure-47 show the options available for each of the parameters.

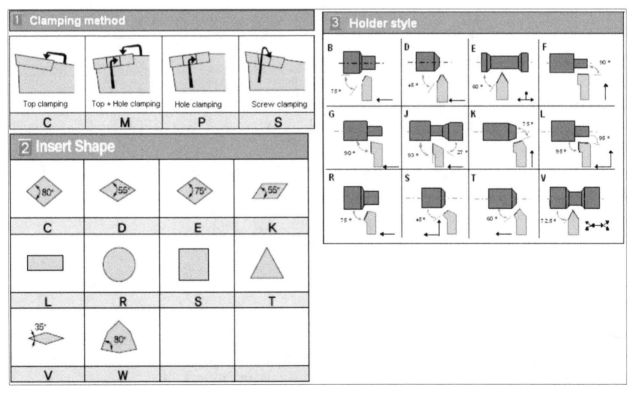

Figure-46. Clamping Method, Insert Shapes, and Holder Style

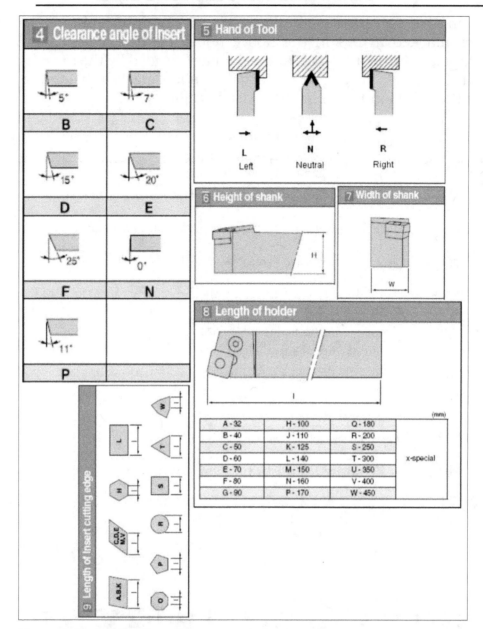

Figure-47. Insert Holder Parameters

CNC Lathe Insert Nomenclature

General CNC Insert name is given as

Meaning of each box in nomenclature is given next.

1 = Turning Insert Shape

The first letter in general turning insert nomenclature tells us about the general turning insert shape, turning inserts shape codes are like C, D, K, R, S, T, V, W. Most of these codes surely express the turning insert shape like

C = C Shape Turning Insert
D = D Shape Turning Insert
K = K Shape Turning Insert

R = Round Turning Insert
S = Square Turning Insert
T = Triangle Turning Insert
V = V Shape Turning Insert
W = W Shape Turning Insert

Figure-48 shows the turning inserts shapes.

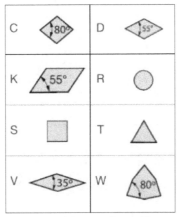

Figure-48. Turning Insert Shapes

The general turning insert shape play a very important role when we choose an insert for machining. Not every turning insert with one shape can be replaced with the other for a machining operation. As C, D, W type turning inserts are normally used for roughing or rough machining.

2 = Turning Insert Clearance Angle

The second letter in general turning insert nomenclature tells us about the turning insert clearance angle.

The clearance angle for a turning insert is shown in Figure-49.

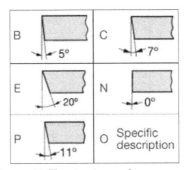

Figure-49. Turning insert clearance angle

Turning insert clearance angle plays a big role while choosing an insert for internal machining or boring small components, because if not properly chosen the insert bottom corner might rub with the component which will give poor machining. On the other hand a turning insert with 0° clearance angle is mostly used for rough machining.

3 = Turning Insert Tolerances

The third letter of general turning insert nomenclature tells us about the turning insert tolerances. Figure-50 shows the tolerance chart.

Code Letter	Cornerpoint (inches)	Thickness (inches)	Inscribed Circle (in)	Cornerpoint (mm)	Thickness (mm)	Inscribed Circle (mm)
A	.0002"	.001"	.001"	.005mm	.025mm	.025mm
C	.0005"	.001"	.001"	.013mm	.025mm	.025mm
E	.001"	.001"	.001"	.025mm	.025mm	.025mm
F	.0002"	.001"	.0005"	.005mm	.025mm	.013mm
G	.001"	.005"	.001"	.025mm	.13mm	.025mm
H	.0005"	.001"	.0005"	.013mm	.025mm	.013mm
J	.002"	.001"	.002-.005"	.005mm	.025mm	.05-.13mm
K	.0005"	.001"	.002-.005"	.013mm	.025mm	.05-.13mm
L	.001"	.001"	.002-.005"	.025mm	.025mm	.05-.13mm
M	.002-.005"	.005"	.002-.005"	.05-.13mm	.13mm	.05-.15mm
U	.005-.012"	.005"	.005-.010"	.06-.25mm	.13mm	.08-.25mm

Figure-50. Insert tolerance chart

4 = Turning Insert Type

The fourth letter of general turning insert nomenclature tells us about the turning insert hole shape and chip breaker type; refer to Figure-51.

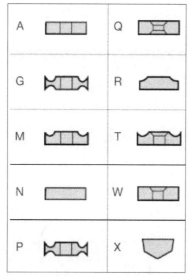

Figure-51. Turning Insert hole shape and chip breaker

5 = Turning Insert Size

This numeric value of general turning insert tells us the cutting edge length of the turning insert; refer to Figure-52.

Figure-52. Turning Insert Cutting Edge Length

6 = Turning Insert Thickness

This numeric value of general turning insert tells us about the thickness of the turning insert.

7 = Turning Insert Nose Radius

This numeric value of general turning insert tells us about the nose radius of the turning insert.

Code	=	Radius Value
04	=	0.4
08	=	0.8
12	=	1.2
16	=	1.6

You can know more about tooling from your tool supplier manual.

DEFINING STOCK FOR PART

The **Stock Manager** tool in **SOLIDWORKS CAM CommandManager** is used to create and manage stock for the part. This stock is the digital model of workpiece on which machining will be performed in the machine shop. The procedure to use this tool is given next.

- Click on the **Stock Manager** tool from the **SOLIDWORKS CAM CommandManager** in the **Ribbon**. The **Stock Manager PropertyManager** will be displayed; refer to Figure-53.

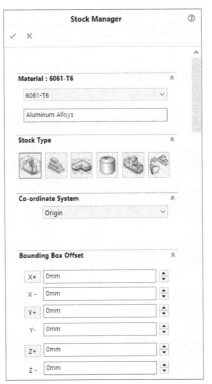

Figure-53. Stock Manager PropertyManager

- Select desired option from the **Material** drop-down to define material for workpiece.
- Select desired button from the **Stock Type** rollout. Various options in this tab are discussed next.

Bounding Box

- By default, the **Bounding Box** button is selected in the dialog box; refer to Figure-53. Hence, a rectangular shaped block of stock is generated.
- Set desired values in edit boxes of **Bounding Box Offset** rollout. If you want to create symmetric workpiece in X direction, then click on the X+ button. Similarly, you can create symmetric workpiece in other directions.
- After setting desired parameters, click on the **OK** button from **PropertyManager** to create the stock.

Pre-defined Bounding Box

- Select the **Pre-defined Bounding Box** button from the **Stock Type** rollout of the **PropertyManager** if you want to specify create a bounding box stock with specified length, width, and thickness.
- Set desired value for offset as discussed earlier.

Extruded Sketch

- Select the **Extruded Sketch** button from the **Stock Type** rollout. The options will be displayed as shown in Figure-54.

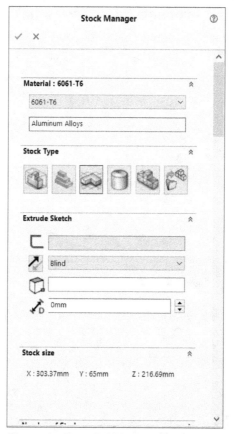

Figure-54. Extruded Sketch stock options

- Select a closed loop sketch to be used for creating an extrude feature. Preview of the extruded workpiece will be displayed; refer to Figure-55.

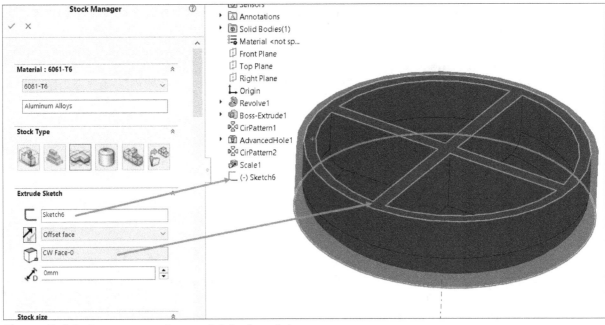

Figure-55. Selecting parameters for extruded sketch workpiece

- Set desired parameters in the **Extruded Sketch** rollout.
- Click on the **OK** button from the **PropertyManager**. The workpiece stock will be created.

Cylindrical Stock

- Click on the **Cylindrical** button from the **Stock Type** rollout to create a cylindrical stock with specified diameter, length, axis location, and so on. Note that this stock is useful for cylindrical models.

STL File

- Select the **STL File** button from the **Stock Type** rollout. The options will be displayed as shown in Figure-56.

Figure-56. STL File stock options

- Click on the **Browse** button from the **STL File** rollout. The **Open** dialog box will be displayed; refer to Figure-57.

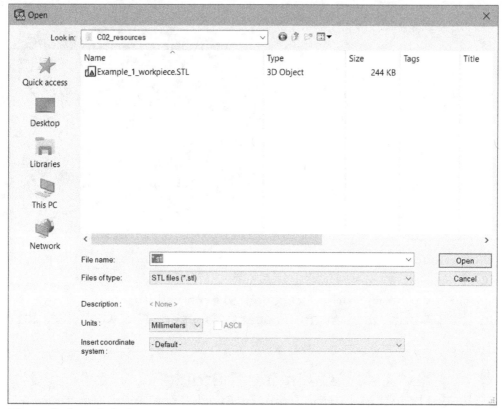

Figure-57. Open dialog box

- Select desired unit from the **Units** drop-down. The model will be imported with selected unit.
- Select desired option from the **Insert coordinate system** drop-down if you want to include a specific coordinate system.
- Select desired STL file from the dialog box and click on the **Open** button. The model will be imported. Note that the model will be placed at default/specified coordinate system.
- Click on the **OK** button to create the stock.

Part File

- Select the **Part File** button from the **Stock Type** rollout. The options will be displayed as shown in Figure-58.

Figure-58. Part File stock options

- Select the **Select Part** radio button if you want to use SolidWorks part as workpiece. Click on the **Browse** button next to field and select desired file.
- Select the **Current Part** radio button if you want to select a configuration of current model as workpiece. Note that this option will be very useful when the workpiece is scaled up configuration of the current model. After selecting this radio button, select desired configuration from the drop-down and set the other parameters; refer to Figure-59.

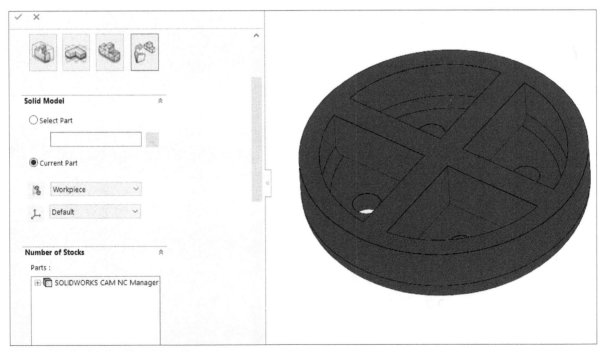

Figure-59. Stock using Part file configuration

- After setting desired parameters, click on the **OK** button to create the stock.

SELF ASSESSMENT

Q1. What is the function of Tool Spindle in 5 axis VMC?

Q2. What is the function of Tool Turret in a machine?

Q3. What is the function of tail stock in lathe machine?

Q4. The options in the .... tab of **Machine** dialog box are used to setup cutting tools to be used for machining.

Q5. Discuss the use of End Mill cutting tool.

Q6. Discuss the use of Dovetail cutter in milling.

Q7. Discuss the use of Engrave mill cutting tool.

Q8. Discuss the use of Barrel mill cutting tool.

Q9. Discuss the nomenclature of CNC lathe insert holder and cutting tool insert.

Chapter 3

Milling Setup and Features

Topics Covered

The major topics covered in this chapter are:

- *Introduction.*
- *Milling Setup.*
- *Extracting Machinable Features.*
- *SolidWorks CAM Options.*
- *Editing Recognized Features.*
- *Creating Mill Features Manually.*
- *Creating Multi Surface Features.*
- *Defining Default Operation Strategies.*
- *Generating Operations Automatically.*

INTRODUCTION

In the previous chapter, you learned about the machine definition and workpiece setup. You learned about creating new machine or modify existing machine to match your physical machine in the workshop. You also learned to create stock material for given part which will be machined by the cutting tool. In this chapter, you will setup milling operations for machining. The tools for machine setup and operation are discussed next.

MILLING SETUP

The **Mill Setup** tool is used to setup cutting direction and machining strategies for given part. The procedure to use this tool is given next.

- Click on the **Mill Setup** tool from the **SOLIDWORKS CAM TBM CommandManager** of the **Ribbon**. The **Mill Setup PropertyManager** will be displayed; refer to Figure-1.

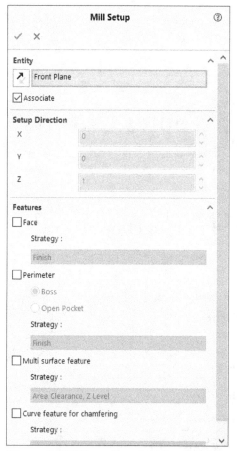

Figure-1. Mill Setup PropertyManager

- Click in the **Entity** selection box of **PropertyManager** and select the face perpendicular to which the machining will be performed. A red mark will be displayed on the model defining the tool direction; refer to Figure-2.
- Select the **Associative** check box from the **Entity** rollout if you want the direction to be associated with selected face. In this way, the change in selected face will be reflected in tool direction.

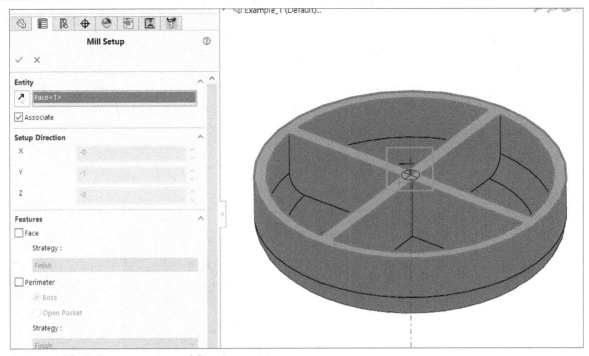

Figure-2. Mark showing cutting tool direction

- Select the **Face** check box from the **Features** rollout if you want to create facing operation. After selecting this check box, select desired option from the **Strategy** drop-down. Select the **Finish** option if you want to perform finishing operation on the face. Select the **Coarse** option if you want to perform roughing operation of the face. Select the **Fine** option if you want to machine for better finishing. Select the **Finish - Islands** option if you want to machine faces of boss features in the part.

- Select the **Blind** toggle button adjacent to **Strategy** drop-down if you want to perform facing operation up to specified depth. De-select this toggle button if you want to perform through facing operation.

- Select the **Perimeter** check box if you want to remove material around the perimeter of part or boss features in the part. A perimeter feature will be added in the setup. Select the **Boss** radio button if you want to use boss features in the model for creating perimeter CAM feature. Select the **Open Pocket** radio button if you want to create perimeter CAM feature around the walls of part. You can set the options in **Strategy** drop-down in the same way as discussed earlier. Select the **Finish -EdgeBreak** option if you want to chamfer irregular edges which are non-planar. If the **Open Pocket** radio button is selected then you can also select **Rough(VoluMill) - Finish** option from the **Strategy** drop-down. VoluMill is a proven toolpath strategy that enables you to cut much faster and deeper than with typical toolpath strategies. At the same time, by avoiding sharp directional changes and controlling the rate of material removal, it can significantly extend tool life. It runs seamless inside of CAM Express, Siemens NX CAM, TopSolid, CAMWorks, NCCS, Cimatron, and GibbsCAM.

- Select the **Multi surface feature** check box from the **Features** rollout if you want to create a machining feature to remove material from different shapes bound the surfaces of part, this includes pockets, areas, and so on. Select the **Area Clearance, Z Level** option from the respective **Strategy** drop-down if you want to perform area clearance machining at specified Z level in the workpiece. The other options in the **Strategy** drop-down are same as discussed earlier.

- Select the **Curve feature for chamfering** check box if you want to create curve features automatically for chamfers in the model. Select desired option from the respective **Strategy** drop-down. The options in this drop-down have been discussed earlier.

- After setting desired parameters, click on the **OK** button from the **PropertyManager**. The features will be created under **Mill Part Setup** node; refer to Figure-3.

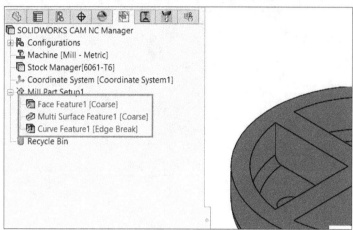

Figure-3. Features created in mill part setup

EXTRACTING MACHINABLE FEATURES AUTOMATICALLY

The **Extract Machinable Features** tool is used to automatically identify the features to be machined in the model. The procedure to use this tool is given next.

- Click on the **Extract Machinable Features** tool from the **SOLIDWORKS CAM CommandManager** in the **Ribbon**. The **SOLIDWORKS CAM Message Window** dialog box will be displayed; refer to Figure-4.

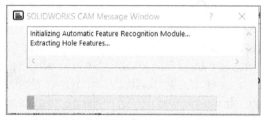

Figure-4. SOLIDWORKS CAM Message Window dialog box

- After automatic identification, the features will be added in the setup as shown in Figure-3.

Note that the features that can be recognized by CAMWorks can be defined in **Options** dialog box. The procedure to define options is discussed next.

SOLIDWORKS CAM OPTIONS

The **SOLIDWORKS CAM Options** tool is used to define basic parameters of SolidWorks CAM. The procedure is given next.

- Click on the **SOLIDWORKS CAM Options** tool from the expanded **SOLIDWORKS CAM CommandManager** in the **Ribbon**; refer to Figure-5. The **Options** dialog box will be displayed as shown in Figure-6.

Figure-5. SOLIDWORKS CAM Options tool

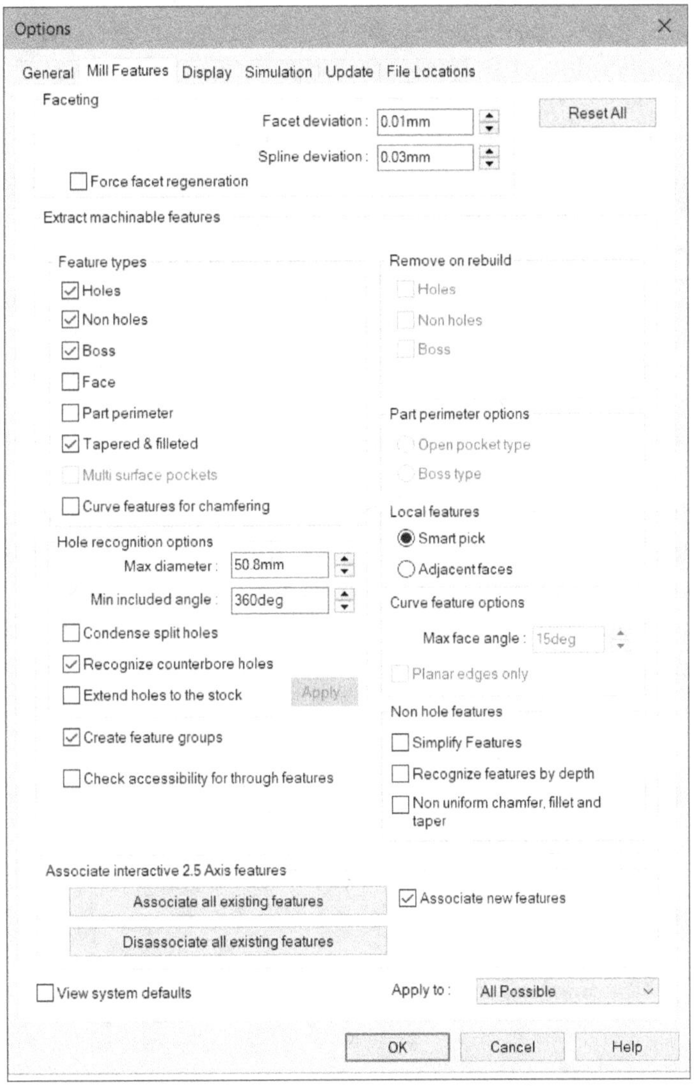

Figure-6. Options dialog box

Mill Features Tab

The options in the **Mill Features** tab are used to define parameters for facet regeneration, machinable feature extraction, and so on. The options in this tab are discussed next.

- Select the **Force facet recognition** check box if you want to regenerate facets of the model during 3 Axis milling up t o specified Facet deviation.

- Set desired value in **Facet deviation** and **Spline deviation** edit boxes of the **Faceting** area of the dialog box to define tolerance within which surface/curve can deviate from original shape. Note that a wrong value specified in these edit boxes can make significant difference between desired part and machined part.
- Select desired check boxes from the **Feature types** area to define which features are to be recognized automatically by **Extract Machinable Features** tool.
- Select desired check boxes from the **Remove on rebuild** area of the dialog box to remove respective features while rebuilding the model for machining. Note that if the features in this area are selected in **Feature types** area then they will not be active in **Remove on rebuild** area.
- If you have selected the **Part perimeter** check box from the **Feature type** area then options in the **Part perimeter options** area of the dialog box will be active. Select desired option from this area. The options in this area have been discussed earlier.
- Select desired option from the **Local features** area of the dialog box to define how local features will be recognized based on user-selected faces. There are two radio buttons in this area; **Smart pick** and **Adjacent faces**. If **Smart pick** option is selected and a face within the intended feature is picked, then Local Feature Recognition (LFR) automatically determines which faces are relevant to create the feature. Multiple faces can be selected to define multiple features. If **Adjacent faces** option is selected and a face is picked, then Local Feature Recognition (LFR) automatically selects faces adjacent to that face and creates features.
- The options in the **Curve feature options** area are available only when **Curve features for chamfering** check box is selected in the **Feature type** area of the dialog box. Set desired value of angle in the **Max face angle** edit box of the **Curve feature options** area of the dialog box to define which edges are to be used for chamfering. If a face has angle more than value specified in the **Max face angle** edit box with respect to horizontal or vertical plane then its edges will not be used for curve chamfering feature; refer to Figure-7.

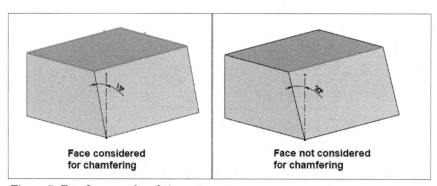

Figure-7. Face for curve chamfering

- Select the **Planar edges only** check box if you want to chamfer only those faces which are flat and parallel to machining plane.
- Specify desired parameters in the **Hole recognition options** area to define the maximum diameter and angle to be used for identifying holes. Specify desired values in **Max diameter** and **Min included angle** edit boxes.
- Select the **Condense split holes** check box if you want Automatic Feature recognition to consider two or more cylindrical faces that are collinear, have the same diameter and would normally be machined together as a single hole.
- Select the **Extend holes to the stock** check box if you want to extend recognized holes upto the stock in both upward and downward directions.

- Select the **Create feature groups** check box if you want to create groups of hole features and non-hole features.
- Select the **Check accessibility for through features** check box if you think there are features in the model that can not be machined directly and can cause tool accessibility problem. In such cases, software will automatically move those features to mill part setup and apply best possible cutting tools and cutting operations.
- Select the **Simplify Features** check box when the **Non hole features** check box is selected in the **Feature types** area of the dialog box if you want to recognize irregular shaped features as combination of small regular shaped features.
- Select the **Recognize features by depth** check box if you want to recognize two connected features of different depth as two separate features.
- Select the **Non uniform chamfer, fillet and taper** check box to identify slots, pockets and other features with non uniform chamfers, fillets, and taper faces.
- Click on the **Associate all existing features** button to link all the machining features with their respected references (faces/axes/points).
- Click on the **Disassociate all existing features** button to break the links between machining features and their respective references.
- Select the **Associate new features** check box to automatically associate new machining features with their respective references.
- Select desired option from the **Apply to** drop-down to define the scope of applying specified settings.
- Select the **View system defaults** check box to check default settings in the dialog box.

General Tab

The options in the **General** tab are used to define general parameters for SOLIDWORKS CAM. These options are discussed next.

- Select the **Save/Restore part** check box if you want to save the part to be machined in SolidWorks CAM project directory.
- Select the **Save/Restore assembly** check box if you want to save the assembly files in SolidWorks CAM project directory.
- Select the **Message window** check box to display progress of SolidWorks CAM machining in message windows.
- Select the **Arrow key navigation** check box to use arrow keys for navigating in **SOLIDWORKS CAM Machinable Features** and **SOLIDWORKS CAM Operation** trees.
- Select the **Dynamic highlight** check box if you want to highlight the feature when cursor moves over it in the Feature or Operation trees.
- Select the **Disable Auto Saving** check box to disable auto saving of project files at auto save time.
- Select the **Show Save Dialog** check box from the **NC Node Lock** area to display **Save** dialog box when part is locked in **SOLIDWORKS CAM NC Manager**.
- Select the **Use SOLIDWORKS CAM configurations** check box if you want to use SolidWorks CAM configuration of model for machining.
- Select the **Update tool selection** check box from the **Technology Database** area of the dialog box to automatically update selected tools based on technology database.
- Select the **Modify tool** radio button from the **Tool library change** area to automatically modify existing tools of project in database when project is saved.

- Select the **Add new tool** radio button to add modified tools of project as new tools in database.
- Select the **Show Options before generation** check box to display options for generating XML files of setup sheets.
- Select the **Omit mill holder message** check box to turn off display of messages that display when updating the holder while changing mill tools.

Display Tab

The options in the **Display** tab are used to set parameters for displaying various entities of model; refer to Figure-8. These options are discussed next.

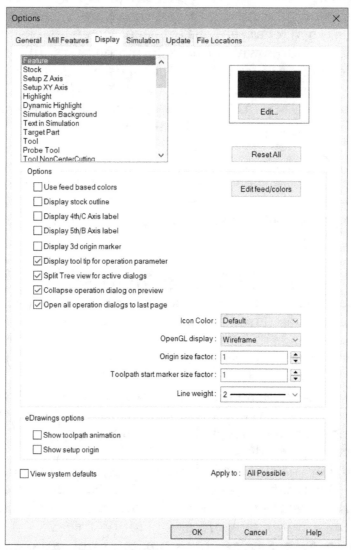

Figure-8. Display tab

- Select desired option from the list box and click on the **Edit** button to change the color of respective interface object.
- Click on the **Reset All** button to reset all the interface objects to original.
- Select the **Use feed based colors** check box to use specified colors for feed representing lines. Click on the **Edit feed/colors** button to modify colors. The **Feed Color Settings** dialog box will be displayed; refer to Figure-9. Set desired range of feed and corresponding color in the dialog box. If you want to add new range of feed color then click on the **Add** button and set other parameters as desired. Click on the **OK** button to apply the settings.

Figure-9. Feed Color Settings dialog box

- Select the **Display stock outline** check box to display stock outlines even when stock is not selected in the Tree.
- Select the **Display 4th/C Axis label** check box to display label of 4th/C axis of machine.
- Select the **Display 5th/B Axis label** check box to display label of 5th/B axis of machine.
- Select the **Display 3d origin maker** check box to display 3D origin with axes.
- Select the **Display tool tip for operation parameter** check box to display operation parameters when an operation is selected in the tree.
- Select the **Split Tree view for active dialogs** check box to divide the main tree as per the options selected.
- Select the **Collapse operation dialog on preview** check box if you want the dialog boxes to collapse automatically when the **Preview** button is selected.
- Select the **Open all operation dialogs to last page** check box if you want all operation dialog boxes open to the active tab of the last closed dialog box.
- Set desired parameters in **Icon Color**, **OpenGL display**, **Origin size factor**, **Toolpath start marker size factor**, and **Line weight** drop-downs.
- Select desired parameters in the **eDrawings** options area of the dialog box. Select the **Show toolpath animation** check box if you want to display animation tab in **eDrawings Manager**. Select the **Show setup origin** check box to display origin in eDrawing.

Simulation Tab

The options in the **Simulation** tab are used to manage parameters related to machining simulation; refer to Figure-10. These options are discussed next.

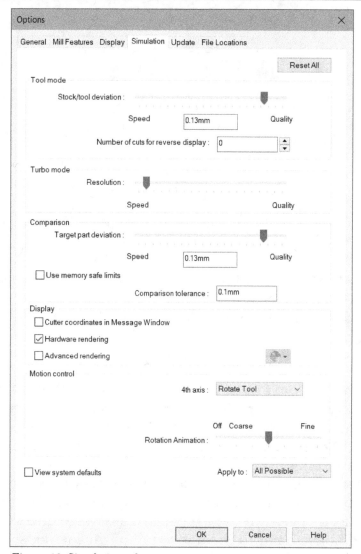

Figure-10. Simulation tab

- Specify desired value of deviation of stock/tool in simulation by using the **Stock/ tool deviation** slider in the **Tool mode** area of the dialog box.
- Specify desired value in **Number of cuts for reverse display** edit box to specify the maximum number of cuts where material can be added to the stock while in Reverse step mode. Note that when using this option, Tool mode simulation will be slower by about 15%. As the number of allowed steps to add material is increased, more system memory is also required.
- Set desired value for the **Resolution** slider of **Turbo mode** area to specify number of facets to be used to represent surface of model during simulation.
- Set desired value for the **Target part deviation** slider in the **Comparison** area of the dialog box to define maximum deviation of simulation model with respect to original model.
- Select the **Use memory safe limits** check box to make sure you do not run out of memory when deviation is set to lower value.
- Set desired value in **Comparison tolerance** edit box to specify tolerance for comparing simulation model and original model.
- Select the **Cutter coordinates in Message Window** check box to display coordinates of cutting tool during simulation.
- Select the **Hardware rendering** check box to improve graphics and quality of simulation using the graphic card.

- Select the **Advanced rendering** check box to display realistic rendering of toolpath simulation by using metallic textures. Click on the drop-down next to **Advanced rendering** check box to define textures for different part of simulation; refer to Figure-11.

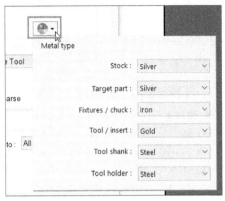

Figure-11. Texture drop-down

- Select desired option from the **4th axis** drop-down to define whether tool is rotating or stock is rotating member of simulation.
- Set the **Rotation Animation** slider to desired position in **Motion control** area to define quality of rotation during simulation.

Update Tab

The options in this tab are used to define features to be automatically updated; refer to Figure-12. These options are discussed next.

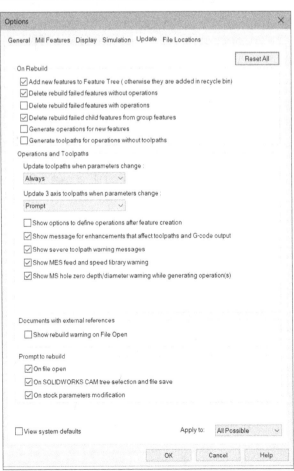

Figure-12. Update tab

- Select the **Add new features to Feature Tree (otherwise they are added in recycle bin)** check box to add new features automatically to Feature Tree on rebuilding.
- Select the **Delete rebuild failed features without operations** check box to delete the machining features which have failed during rebuilding and do not have associated operations.
- Select the **Delete rebuild failed features with operations** check box to delete the machining features which have failed during rebuilding and have associated operations.
- Select the **Delete rebuild failed child features from group features** check box to delete the child features of feature groups which have failed during rebuilding.
- Select the **Generate operations for new features** check box to automatically generate operations for new machining features.
- Select the **Generate toolpaths for operations without toolpaths** check box to automatically generate toolpaths for operations which do not have toolpaths associated with them.
- Select desired option from the **Update toolpaths when parameters change** drop-down of **Operations and Toolpaths** area in the dialog box. Select the **Always** option if you want to always update toolpaths when parameters change. Select the **Prompt** option if you want system to prompt when there is change in parameters of toolpath for updating. Select the **Never** option if you do not want to automatically update the toolpath after changing parameters.
- Select desired option from the **Update 3 axis toolpaths when parameters change** drop-down to define how toolpaths will be updated when parameters are changed.
- Select the **Show options to define operations after feature creation** check box to automatically display options related to operations for created feature.
- You can set the other options in the **Operations and Toolpaths** area of the dialog box in the same way.

File Locations Tab

The options in this tab are used to define locations of various SolidWorks CAM related files and directories; refer to Figure-13. The options in this tab are discussed next.

- The options in the **SOLIDWORKS CAM data folder** and **TechDB Location** fields can not be edited as these parameters are set during installation.
- Click on the **Browse** button for **Setup sheet images folder** edit box and set the location where you want to save the setup sheet images.
- Select the **Open G-code file in SOLIDWORKS CAM NC Editor** check box to open the G-codes in SOLIDWORKS CAM NC Editor application after generating codes.
- Select the **Open G-code file in following application** check box and select desired application using the **Browse** button.
- Select the **Open setup (.set) file in following application** check box to define the application to be used for opening setup file of SolidWorks CAM.

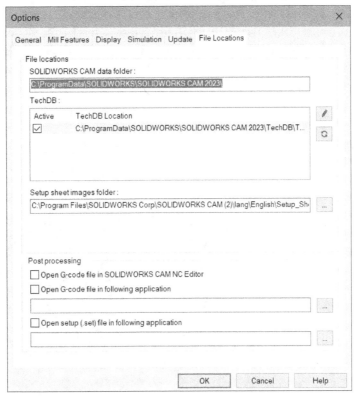

Figure-13. File Locations tab

After setting desired parameters, click on the **OK** button.

EDITING RECOGNIZED FACE FEATURES

The automatically recognized features are displayed in **Mill Part Setup** node of **SOLIDWORKS CAM Feature Tree**. The procedure to modify recognized features is given next.

• Right-click on the feature that you want to modify from the **SOLIDWORKS CAM Feature Tree**. A shortcut menu will be displayed; refer to Figure-14.

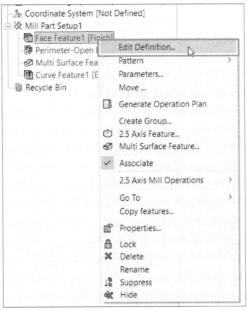

Figure-14. Shortcut menu option

- Click on the **Edit Definition** option from the shortcut menu. Since most of the features in SolidWorks CAM are 2.5 Axis features so, the **2.5 Axis Feature: Select Entities PropertyManager** will be displayed; refer to Figure-15 (for Face feature).

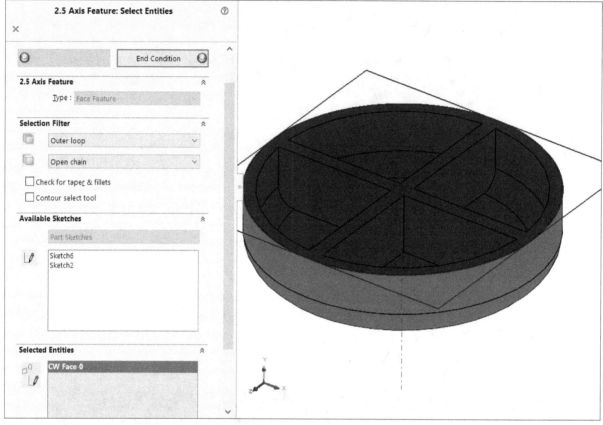

Figure-15. 2.5 Axis Feature Select Entities PropertyManager

- Click in the **Selected Entities** selection box and select the face on which you want to perform facing. To remove an entity from selection box, right-click on it and select the **Delete** option from the shortcut menu; refer to Figure-16.

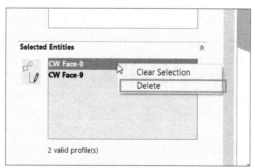

Figure-16. Deleting selected feature

- Note that all the sketches used in model in SolidWorks are displayed in the **Available Sketches** selection box. You can select a sketch for defining the boundary of facing operation; refer to Figure-17.

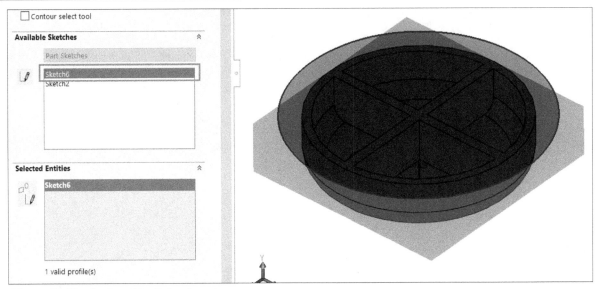

Figure-17. Sketch selected for facing boundary

- Select the **Contour select tool** check box to select multiple entities for defining contour for cutting. For features like open pocket and face features, you need to select one closed entity for defining contour. For features that require open chain like Curve, Open Profile, Engrave feature etc. you can select one or more features to define contour.

- Select the **Check for taper & fillets** check box to automatically select tapers and fillets of the selected face(s).

Defining End Condition

- Click on the **End Condition** button from the **PropertyManager**. The **2.5 Axis Feature: End Conditions PropertyManager** will be displayed; refer to Figure-18.
- Select desired option from the **Strategy** drop-down. You can select **Finish**, **Coarse**, **Fine**, or **Finish - Islands** strategy.
- Select desired option from the **End Condition - 1** drop-down in the **End condition - Direction 1** rollout of **PropertyManager**. There are six options in this drop-down; **Blind**, **Upto Face**, **Offset From Face**, **Upto Vertex**, **Upto Stock**, and **Upto Ref Plane**. Select the **Blind** option and specify desired depth value for facing operation; refer to Figure-19. Select the **Upto Face** option from the drop-down and select the face upto which you want to perform facing. Select the **Offset From Face** option from the drop-down and set desired value of offset from selected face upto which you want to perform facing. Select the **Upto Vertex** option from the drop-down and select a vertex upto which you want to perform facing operation. Select the **Upto Stock** option from the drop-down if you want to perform facing operation upto stock limit. Select the **Upto Ref Plane** option from the drop-down if you want to perform facing operation upto selected reference plane and select desired plane to define depth. Click on the **Reverse direction** button adjacent to **End Condition - 1** drop-down to reverse depth direction.

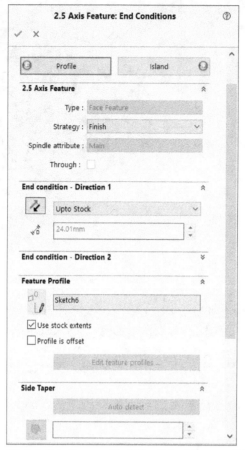

Figure-18. 2.5 Axis Feature: End Conditions PropertyManager

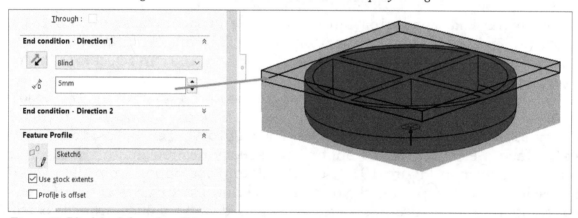

Figure-19. Blind depth for end condition

- Select the **Use stock extents** check box to expand the face machining area upto stock boundaries; refer to Figure-20.

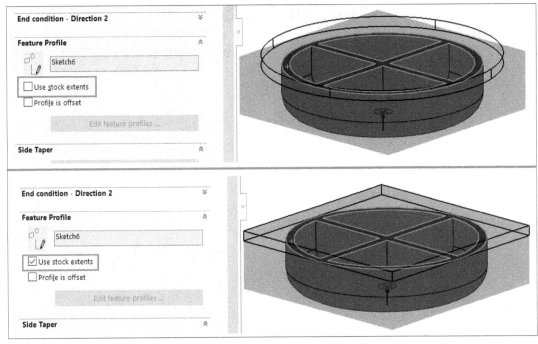

Figure-20. Stock extent options

- When corner slot or slot on side of part is machined then profile is offset automatically by tool diameter value beyond the stock boundary so that finish is achieved edges. Select the **Profile is offset** check box if the cutting profile already including offset value for slots.

Moving Feature

- Click on the **Move** button from the **Move** rollout at the bottom in the **PropertyManager**. The **Move Feature PropertyManager** will be displayed; refer to Figure-21.

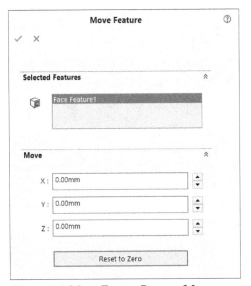

Figure-21. Move Feature PropertyManager

- Click on the **Reset to Zero** button to automatically move feature at origin. Using the X, Y, and Z sliders to move the feature at desired location.

Island Entities

- Click on the **Island** button from the **PropertyManager**. The **2.5 Axis Feature: Island Entities PropertyManager** will be displayed; refer to Figure-22.

Figure-22. 2.5 Axis Feature Island Entities PropertyManager

- Click on the **Auto detect** button from the **PropertyManager**. The islands will be detected automatically in the model.
- Set desired options in the **Selection Filter** rollout to define selection criteria for selecting islands in the model. Islands are the boss features in the model.
- Select the **Contour select tool** check box to use contours for defining islands.
- After setting desired parameters, click on the entities to select for island.
- Click on the **OK** button from the **PropertyManager** to modify/create the feature.

CREATING MILL FEATURES MANUALLY

Various tools to create mill features are available in the **Tools > SOLIDWORKS CAM > New > Feature** menu; refer to Figure-23. The options in this cascading menu are active when **Mill Part Setup** node is selected in the **SOLIDWORKS CAM Feature Tree**. The tools in this drop-down are discussed next.

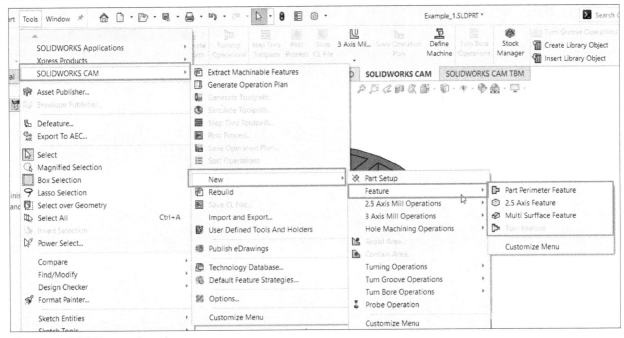

Figure-23. Mill Feature drop-down

CREATING PART PERIMETER FEATURE

The **Part Perimeter Feature** tool in **Feature** cascading menu is used to machine outer boundary of the part from the stock. The procedure to use this tool is given next.

- Select the **Mill Part Setup** node from the **SOLIDWORKS CAM Feature Tree** and click on the **Part Perimeter Feature** tool from the **Tools > SOLIDWORKS CAM > New > Feature** menu. The **New perimeter feature PropertyManager** will be displayed; refer to Figure-24.

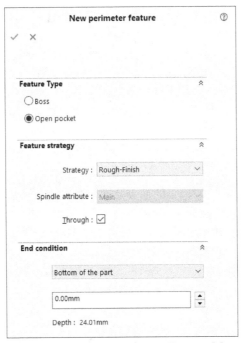

Figure-24. New perimeter feature PropertyManager

- Select the **Boss** radio button from the **Feature Type** rollout if you want to create perimeter feature around boundaries of boss features in the part. Select the **Open pocket** radio button from the rollout if you want to include the open pocket boundaries along with boss feature to create machine boundaries; refer to Figure-25.

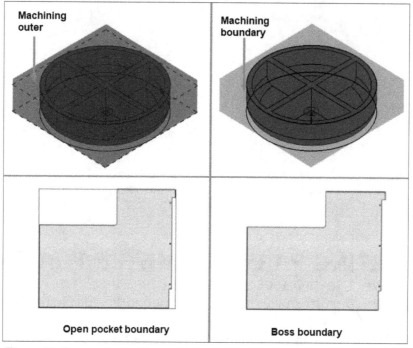

Figure-25. Open pocket and boss perimeter features

- Select desired option from the **Strategy** drop-down in the **Feature strategy** rollout to define machining strategy. The options in this drop-down have been discussed earlier.
- Select desired option from the **Spindle attribute** drop-down to define spindle axis if the machine have multiple spindles.
- Select the **Through** check box if you want the perimeter feature to be through. If this check box is not selected then the features will be created with blind option automatically so the feature will be create upto specified depth.
- Set desired options in the **End condition** rollout as discussed earlier to define depth of feature.
- Click on the **OK** button from the **PropertyManager** to create the feature.

CREATING 2.5 AXIS FEATURE

The 2.5 axis milling operations have restriction of cutting only along 2 axis simultaneously while leaving level along 3rd axis fixed. For example, the cutting tool can move on XY plane but movement along Z axis will be zero. Note that the cutting tool will be free to move along all three axis during non-cutting passes. In SolidWorks CAM, we can create 2.5 axis features which will later be used to generate toolpaths. The procedure to create 2.5 axis features is given next.

- Click on the **2.5 Axis Feature** tool from **Tools > SOLIDWORKS CAM > New > Feature** menu. The **2.5 Axis Feature: Select Entities PropertyManager** will be displayed; refer to Figure-26.
- Select desired option from the **Type** drop-down to define cutting strategy. Examples for selecting features are shown in Figure-27 and Figure-28.

Figure-26. 2.5 Axis Feature Select Entities PropertyManager

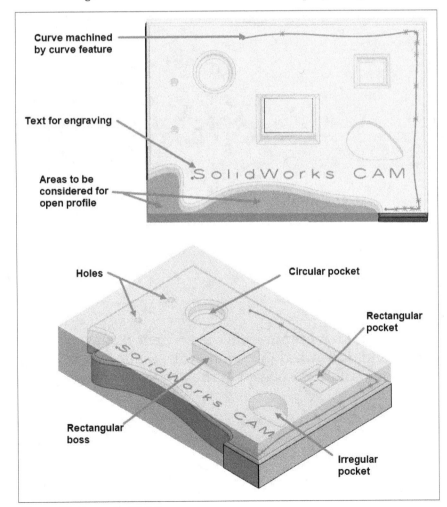

Figure-27. Example of 2.5 features

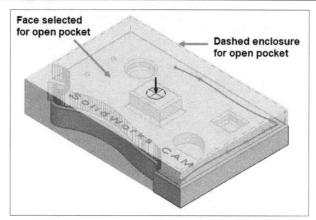

Figure-28. Example of 2.5 axis open pocket feature

The procedures to create these features are discussed next.

Creating Face Feature

- Select the **Face Feature** option from the **Type** drop-down in **PropertyManager** to remove extra material from the face of part. This is the first operation performed on workpiece to remove extra material from top face of part.
- By default, **Outer loop** option is selected in the **Face Selection** drop-down of **Selection Filter** rollout. Select the top face of model; refer to Figure-29.

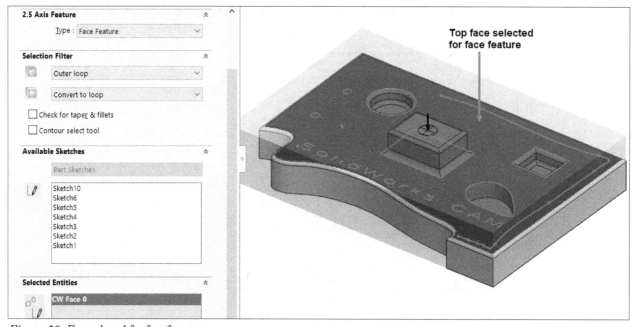

Figure-29. Face selected for face feature

- Click on the **End Condition** button from the **PropertyManager** to define the cutting depth in upward and downward directions with respect to selected face. The end condition operations will be displayed in the **PropertyManager**; refer to Figure-30.

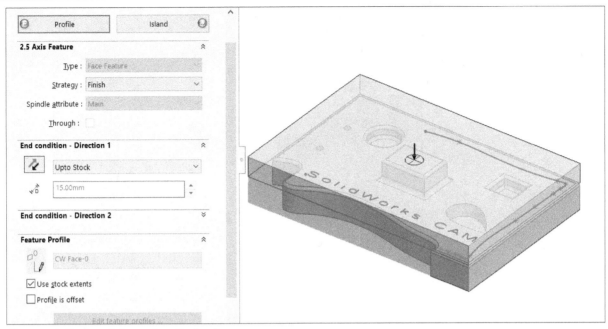

Figure-30. End condition options for face feature

- Select desired option from the **Strategy** drop-down to define what type of machining feature you want to create. Select the **Finish - Islands** option from the **Strategy** drop-down if there are islands on the part face.
- Set desired values in **End condition - Direction 1** and **End condition - Direction 2** rollouts to define scope of facing operation.
- Select the **Use stock extents** check box to remove material from the stock upto its extents rather then using the selected face boundaries; refer to Figure-31.

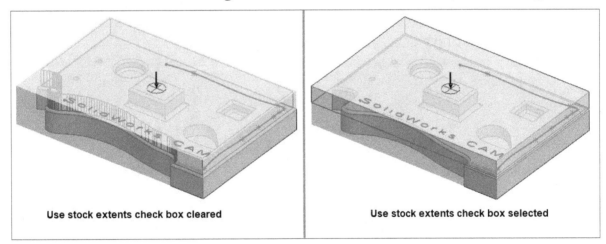

Figure-31. Effect of stock extents

- If your workpiece has tapered walls then select the **Auto detect** button to automatically recognize the taper walls.
- Click on the **Island** button to define faces for islands. The options will be displayed as shown in Figure-32.

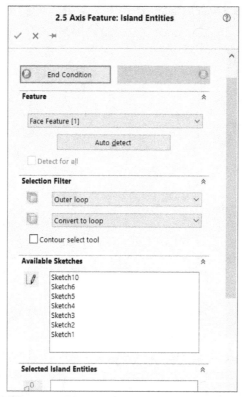

Figure-32. Island entities options

- Click on the **Auto detect** button from the **Feature** rollout to automatically detect the island features. The islands will be selected and displayed in the **Selected Island Entities** selection box. You can remove the entities from selection box as discussed earlier and use the selection filters to select the islands manually; refer to Figure-33. Make sure to remove unnecessary curves from **Selected Island Entities** selection box.

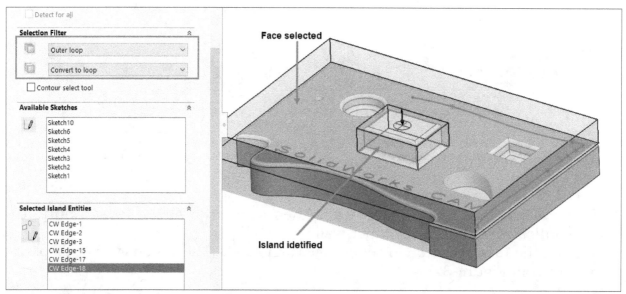

Figure-33. Island identified

- Set the depth/height of island in **End condition** rollout of the **PropertyManager**.
- Click on the **OK** button from the **PropertyManager** to create the feature; refer to Figure-34.

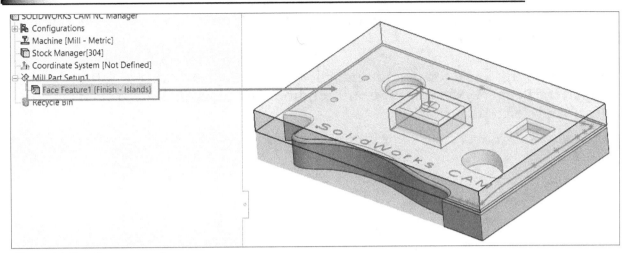

Figure-34. Face feature created

Creating Pocket Feature

• Select the **Pocket** option from the **Type** drop-down in the create different type of closed region pockets (rectangular, circular, and irregular pockets).
• Using the options in **Selection Filter** rollout, select desired loops for pocket features. The options of rollout are discussed next.

Selection Filters

Set desired selection filters in the **Selection Filter** rollout. Select the **Outer loop** option from the **Face Selection** drop-down in the **Selection Filter** rollout to select outer loop of selected face. The outer edge of selected face will be selected and if **Convert to loop** option is selected in the **Edge Selection** drop-down then the outer edges will form a closed loop for machining. Select the **Inner loops** option from the **Face Selection** drop-down to select all closed loops in selected face. All the loops will be selected and displayed in the **Selected Entities** rollout; refer to Figure-35. Select the **Window selection** option from the **Face Selection** drop-down and draw a selection rectangle to collect all the closed loops within the selection.

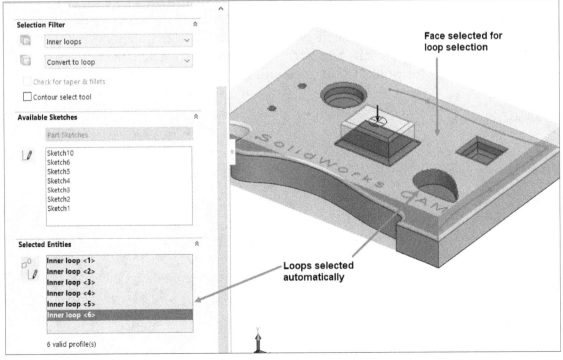

Figure-35. Selecting internal loops

Select the **Open chain** option from the **Edge Selection** drop-down to include open chains during selection.

- After selecting loops for machining, click on the **End Condition** button from the top in the **PropertyManager**. The end condition options will be displayed for pocket feature; refer to Figure-36.

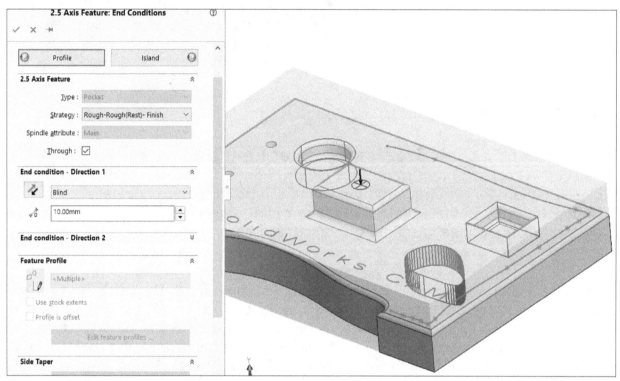

Figure-36. End condition options for pocket feature

- Select desired machining strategy from the **Strategy** drop-down.
- Set the end conditions for both directions as desired in the **End condition - Direction 1** and **End condition - Direction 2** rollouts.
- Click on the **Island** button from the **PropertyManager**. The options for identifying islands in the pocket features will be displayed.
- Select the island faces as needed and click on the **OK** button from the **PropertyManager**. The pocket features will be created; refer to Figure-37.

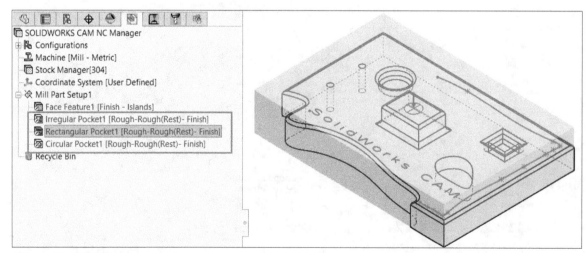

Figure-37. Pocket features created

Creating Slot Feature

- Select the **Slot** option from the **Type** drop-down in the **PropertyManager**. The options to select faces for slots will be displayed.
- Select desired faces for slot features; refer to Figure-38. (Note that slot feature are similar to pocket features but with one side open to edge of workpiece.)

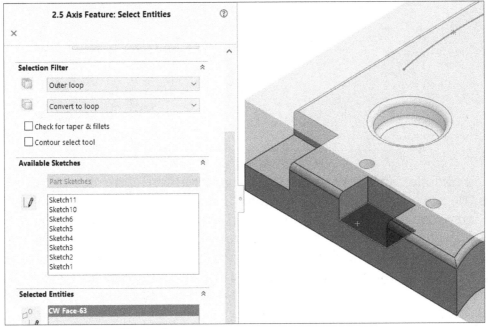

Figure-38. Face selected for slot feature

- Click on the **End Condition** button from the **PropertyManager**. The preview of slot feature will be displayed with options to define depth of slot; refer to Figure-39.

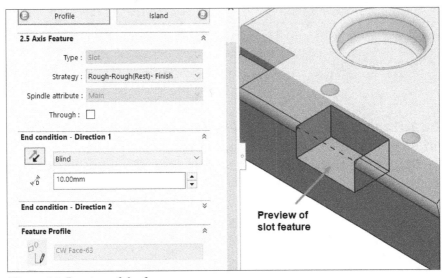

Figure-39. Preview of slot feature

Feature Profile

Note that the feature profile is generated automatically in the **Feature Profile** rollout. If you want to modify the profile then click on the **Edit feature profiles** button from **Feature Profile** rollout. The **Feature Profiles PropertyManager** will be displayed; refer to Figure-40.

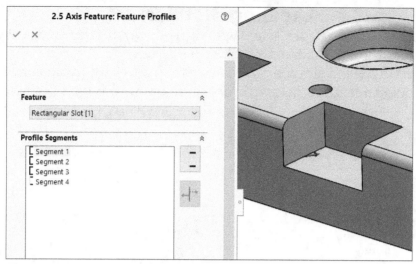

Figure-40. Feature Profiles PropertyManager

For features like slot and corner slot, you can modify the air segment of profile. Air segment is the part of profile which is open and not bound by walls of part. For Open Profile or Curve features, you can change the direction of cutting for the segment. To convert a segment of profile as air segment, select it from the selection box and click on the **Air segment** button; refer to Figure-41. You will learn about open profile and curve feature segments later. After setting profile, click on the **OK** button from the **PropertyManager**. The **2.5 Axis Feature: End Conditions PropertyManager** will be displayed again.

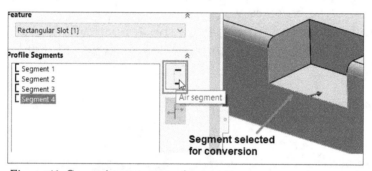

Figure-41. Converting segment to air segment

- Click on the **Island** button from the **PropertyManager** if there are any islands in the slot and select islands as discussed earlier.
- Click on the **OK** button from the **PropertyManager**. The slot feature will be created; refer to Figure-42.

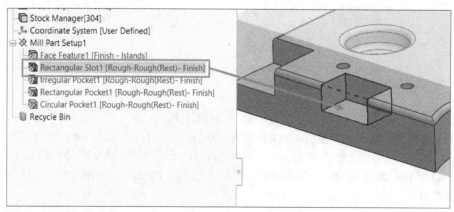

Figure-42. Slot feature created

Creating Corner Slot Features

- Select the **Corner Slot** option from the **Type** drop-down at the top in **PropertyManager**. Select the face to be used for corner slot; refer to Figure-43.

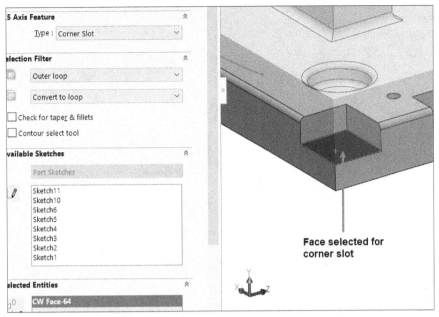

Figure-43. Face selected for corner slot

- Click on the **End Condition** button from the **PropertyManager** to define depth of cut. The options to define parameters for corner slot depth will be displayed with preview of slot; refer to Figure-44.

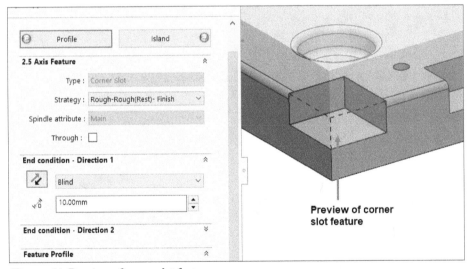

Figure-44. Preview of corner slot feature

- Set desired depth value for slot and strategy for cutting operation. You can modify the feature profile by using **Edit feature profiles** button in the **Feature Profile** rollout as discussed earlier. If there are islands in the slot then click on the **Island** button and set desired parameters.
- Click on the **OK** button from the **PropertyManager** to create the feature.

Creating Boss Features

- Select the **Boss** option from the **Type** drop-down and select faces for boss features using selection filters; refer to Figure-45.

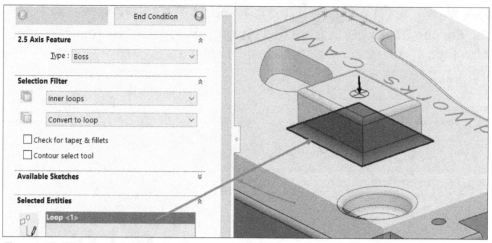

Figure-45. Edge loop selected for boss feature

- Set the parameter as discussed earlier and click on the **OK** button from the **PropertyManager**.

Creating Hole Feature

- Select the **Hole** option from the **Type** drop-down at the top in the **PropertyManager** and select edges of holes for creating hole feature.
- Click on the **End Condition** button from the top in **PropertyManager**. The options to define depth of hole(s) will be displayed with preview of hole feature; refer to Figure-46.

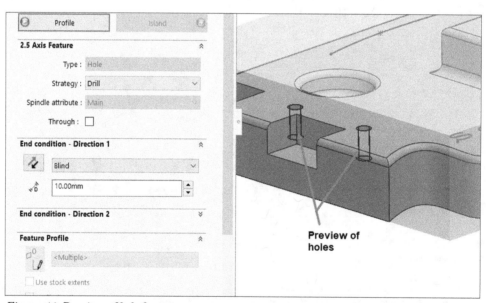

Figure-46. Preview of hole features

- Set desired depth of hole. Select the **Through** check box if the holes are through.
- Click on the **OK** button from the **PropertyManager** to create the hole feature.

Creating Open Pocket Feature

- Select the **Open Pocket** option from the **Type** drop-down in the **PropertyManager** and select the faces/edges for open pocket; refer to Figure-47. Note that open pockets have all sides open for tool movement.

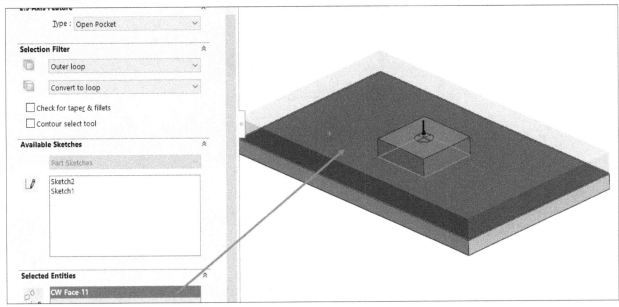

Figure-47. Face selected for open pocket

- Click on the **End Condition** button to set depth of feature. Preview of feature will be displayed; refer to Figure-48.

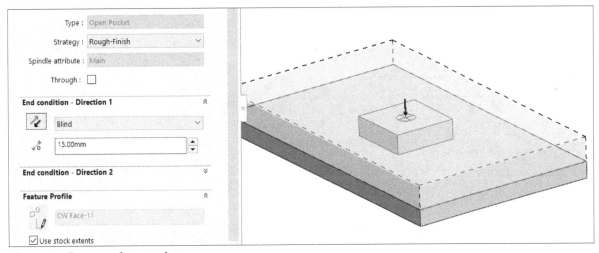

Figure-48. Preview of open pocket

- Set desired parameters for depth and cutting strategy.
- Click on the **Island** button and define island faces if needed.
- Click on the **OK** button to create the feature.

Creating Open Profile Feature

- Select the **Open Profile** option from the **Type** drop-down in the **PropertyManager** to machine side profiles of workpiece which are open. Select desired faces/edges to be used for cutting material around the open profile; refer to Figure-49.

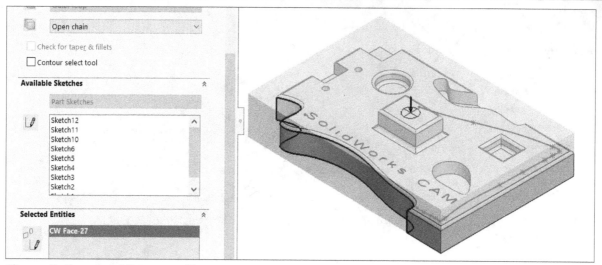

Figure-49. Face selected for open profile

- Click on the **End Condition** button from the top in the **PropertyManager** to define depth of profile.
- Set desired depth value, preview of open profile will be displayed; refer to Figure-50.

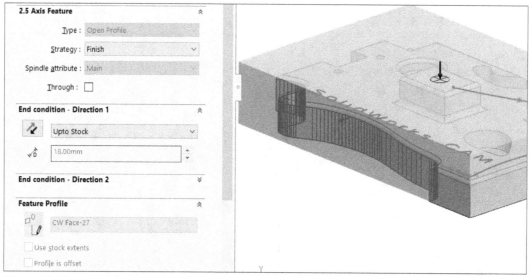

Figure-50. Preview of open profile

- Click on the **Edit feature profiles** button from the **Feature Profile** rollout. The **2.5 Axis Feature: Feature Profiles PropertyManager** will be displayed with options to flip cutting direction for open profile; refer to Figure-51.

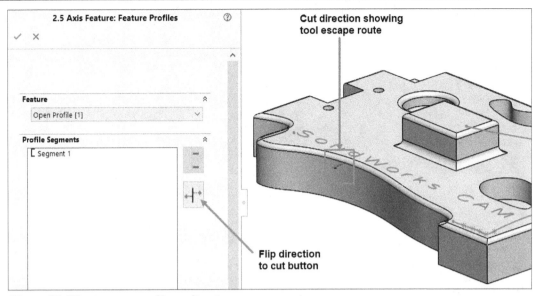

Figure-51. Flipping open profile cut direction

- After setting desired parameters, click on the **OK** button from the **PropertyManager**.

Creating Engrave Feature

- Select the **Engrave Feature** option from the **Type** drop-down and select the sketch for engraving from the **Available Sketches** rollout; refer to Figure-52.

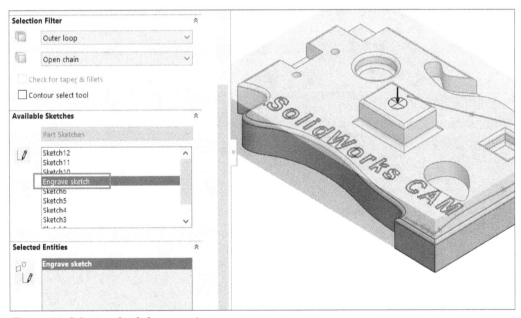

Figure-52. Selecting sketch for engraving

- Click on the **End Condition** button from the **PropertyManager**. The options to define depth will be displayed; refer to Figure-53.

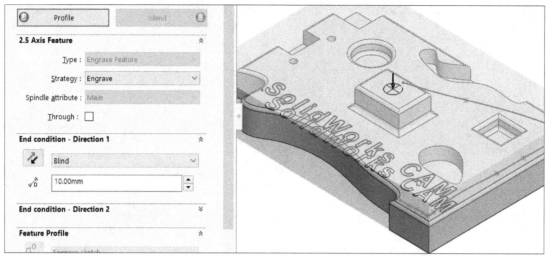

Figure-53. Options to define depth of engraving

- Specify desired depth of engrave feature in the **End condition - Direction 1** rollout.
- Set the other parameters as desired and click on the **OK** button from the **PropertyManager**.

Creating Curve Feature

- Select the **Curve Feature** option from the **Type** drop-down in the **PropertyManager**. The options to select curves for machining features will be displayed; refer to Figure-54.

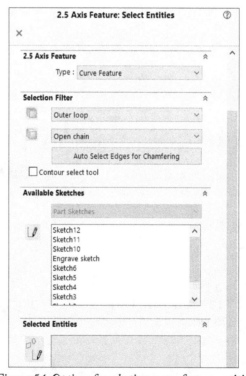

Figure-54. Options for selecting curve feature entities

- Click on the **Auto Select Edges for Chamfering** button from the **Selection Filter** rollout if you want to create chamfer on sharp edges; refer to Figure-55.

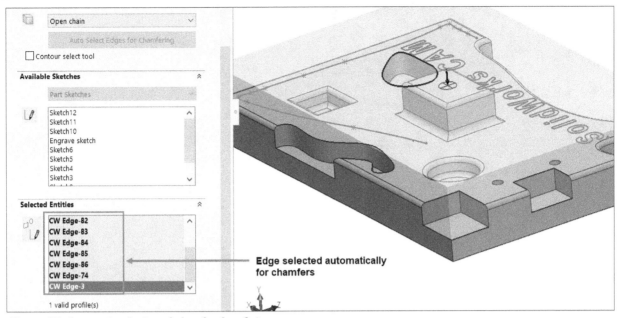

Figure-55. Automatic selection of edges for chamfering

- Click on the **End Condition** button from the **PropertyManager**. Preview of feature will be displayed with options to define depth of curve feature; refer to Figure-56.

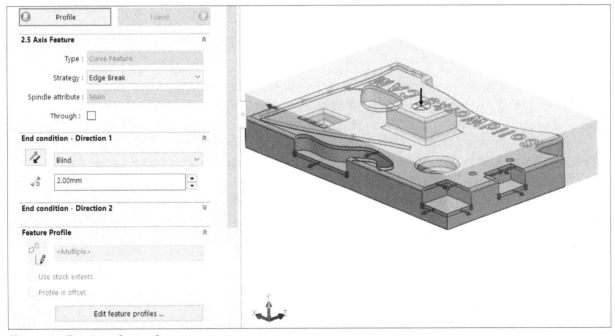

Figure-56. Preview of curve feature

- Set desired strategy, depth, and feature profile for curve feature.
- Click on the **OK** button from the **PropertyManager** to create the feature.

CREATING MULTI SURFACE FEATURE

The **Multi Surface Feature** tool is used to generate 3 axis, 4 axis, and 5 axis toolpaths. In these toolpaths, tool moves along 3, 4, or 5 axis simultaneously. The procedure to use this tool is given next.

- Click on the **Multi Surface Feature** tool from **Tools > SOLIDWORKS CAM > New > Feature** menu. The **Multi Surface Feature PropertyManager** will be displayed; refer to Figure-57.

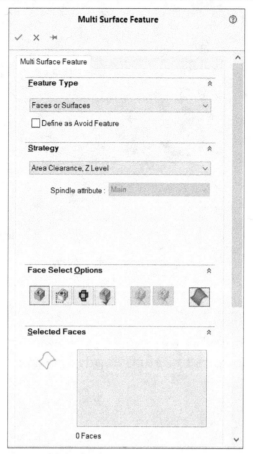

Figure-57. Multi Surface Feature PropertyManager

- Select desired option from the **Feature Type** drop-down to define model features to be selected for multi-surface machining. Select the **Faces or Surfaces** option from the drop-down if you want to use selected faces/surfaces of part for machining.
- Select the **All Displayed** option from the drop-down if all the faces/surfaces of part are to be used for machining.
- Select the **STL File** option from the drop-down if you do not have model created in SolidWorks but have STL file of model for machining. On selecting this option, click on the **Browse** button from the **STL File** rollout; refer to Figure-58 and select desired file for machining.
- Select the **Faces by Color** option from the drop-down to machine faces selected by their assigned colors; refer to Figure-59. After selecting this option, select check boxes for faces to be used for machining.
- Select the **Faces by Surface Finish** option from the drop-down if you want to create machining features based on surface finish value assigned to them. After selecting this option, you need to select check boxes for faces to be used for machining; refer to Figure-60. Note that you can apply surface finish symbol to a face by using the **Surface Finish** tool from the **MBD CommandManager** in the **Ribbon**.

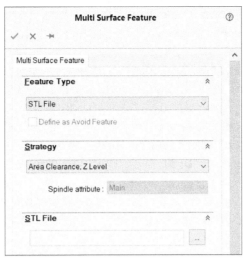

Figure-58. STL file options for features

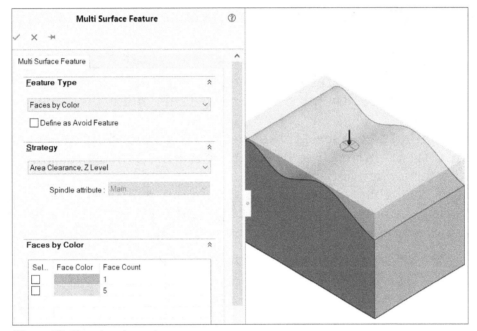

Figure-59. Faces by color options

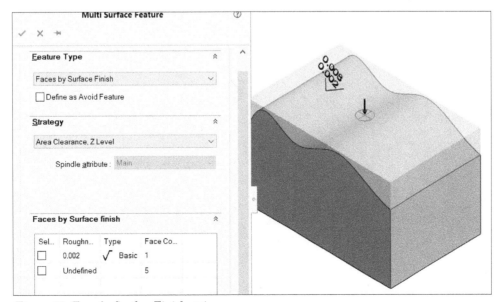

Figure-60. Faces by Surface Finish option

- Select the **Define as Avoid Feature** check box to avoid selected faces while machining. So, when operation plan is generated for features then these faces are not machined.
- Select the **Area Clearance, Z Level** option from the **Strategy** drop-down to machine area clearance and Z level. You can select the **Coarse** or **Fine** option also to define cutting strategy. Note that you can create different cutting strategy using the **Technology Database** options.
- The options of **Face Select Options** rollout are available when **Faces or Surfaces** option is selected in the **Feature Type** drop-down. You can use selection filters in this rollout refine face/surface selection.
- After setting desired parameters in the **PropertyManager**, click on the **OK** button. The feature will be created and added in the **SOLIDWORKS CAM Feature Tree**.

After creating the features, next step is to create operation plan for defining cutting parameters. For different type of features, you can define different type of default operation strategies. The procedure to define default feature strategies is given next.

DEFAULT FEATURE STRATEGIES

The **Default Feature Strategies** tool is used to define default strategy for generating operation plan of features. The procedure to use this tool is given next.

- Click on the **Default Feature Strategies** tool from the **SOLIDWORKS CAM** cascading menu in the **Tools** menu. The **Default Feature Strategies** dialog box will be displayed; refer to Figure-61.
- The features available in the current document will be displayed in boldface. Click in the field under **Default Strategy** column next to the feature for which you want to define strategy. The options will be displayed in a drop-down; refer to Figure-62.

Figure-61. Default Feature Strategies dialog box

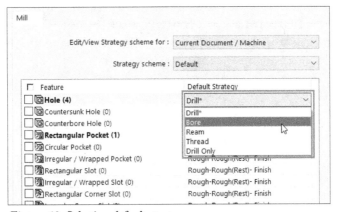

Figure-62. Selecting default strategy

- Select desired options from the drop-downs next to features.
- Click on the **Apply** button to apply changes in strategy.

If you want to change default strategies in Technology Database then select the **Technology Database** option from the **Edit/View Strategy scheme for** drop-down in the top of dialog box. The options will be displayed as shown in Figure-63.

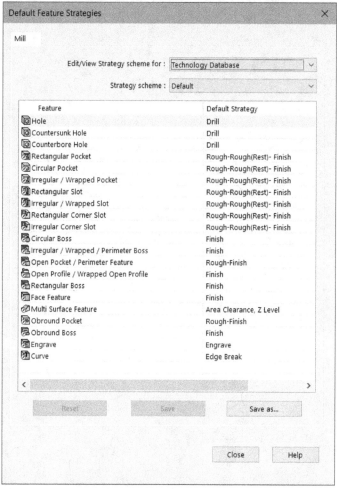

Figure-63. Technology Database options

- Set the default strategies for various features in the same way as discussed earlier. Click on the **Save as** button to save feature strategies. The **Save Default Feature Strategies** dialog box will be displayed; refer to Figure-64.

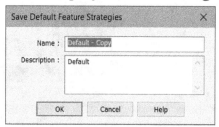

Figure-64. Save Default Feature Strategies dialog box

- Specify desired name and description for strategies list, and click on the **OK** button. The feature strategies list will be saved.
- Click on the **Close** button to exit the dialog box.

GENERATING AUTOMATIC OPERATION PLAN

The **Generate Operation Plan** tool is used to automatically generate operations for recognized machining features based on rules specified in the Technology Database. The procedure to use this tool is given next.

- Click on the **Generate Operation Plan** tool from the **SOLIDWORKS CAM CommandManager** in the **Ribbon**. The **SOLIDWORKS CAM Message Window** will be displayed; refer to Figure-65.

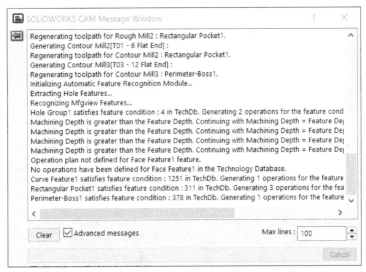

Figure-65. SOLIDWORKS CAM Message Window

- Note that based on features created and parameters defined in the Technology Database, the operation plan is created; refer to Figure-66.

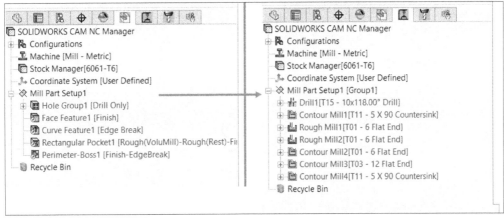

Figure-66. Operation plans generated automatically

Note that you have created the operation plan automatically. You can also create operation plan manually with specified parameters or you can modify parameters of automatically created operation plans. These options are discussed in next chapter.

SELF ASSESSMENT

Q1. The **Mill Setup** tool is used to setup cutting direction and automatic machining strategies for given part. (T/F)

Q2. Select the .... check box from **Mill Setup PropertyManager** if you want to remove material around the perimeter of part or boss features in the part.

Q3. The .... tool is used to automatically identify the features to be machined in the model.

Q4. Select the **Condense split holes** check box if you want to use Automatic Feature recognition to consider two or more cylindrical faces that are collinear, have the same diameter and would normally be machined together as a single hole. (T/F)

Q5. The 2.5 axis milling operations have restriction of cutting only along 2 axis simultaneously while leaving level along 3rd axis fixed. (T/F)

Q6. Air segment is the part of profile which is open and not bound by walls of part. (T/F)

Q7. The **Multi Surface Feature** tool is used to generate 3 axis, 4 axis, and 5 axis toolpaths. (T/F)

Chapter 4

Milling Operations

Topics Covered

The major topics covered in this chapter are:

- *Introduction.*
- *Modifying Area Clearance Operation Plan.*
- *Creating 2.5 Axis Mill Operations Manually.*
- *Creating Hole Machining Operations.*
- *3 Axis Mill Operations.*
- *Probing Operation.*

INTRODUCTION

In previous chapter, you have learned to create features for mill machining. You have also learned to create operation plans automatically for created features. In this chapter, you will learn to modify automatically generated operations and create new operation plans manually.

MODIFYING AREA CLEARANCE OPERATION PLAN

The operations automatically or manually generated for mill features are available in the **Mill Part Setup** node of **SOLIDWORKS CAM Operation Tree**. The procedure to modify an operation is given next.

*   Right-click on an operation to be modified and select the **Edit Definition** option from the shortcut menu; refer to Figure-1. The **Operation Parameters** dialog box will be displayed; refer to Figure-2.

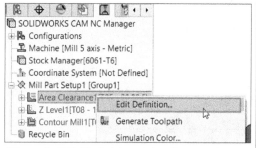

Figure-1. Edit Definition option for operation

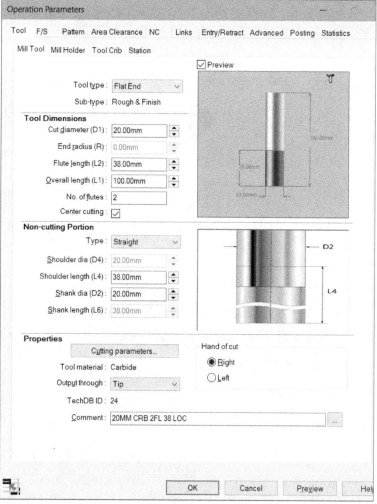

Figure-2. Operation Parameters dialog box

Various parameters in this dialog box are discussed next.

Tool Parameters

The options in the **Tool** tab are used to define parameters related to cutting tool; refer to Figure-2. Various options in this tab are discussed next.

- Set desired option in **Mill Tool** sub-tab of **Tool** tab in the dialog box to define tool type, cutting length, and non-cutting length of the tool. The options in this tab have been discussed earlier.
- Similarly, you can specify parameters related to **Mill Holder**, **Tool Crib**, and **Station** in their respective sub-tabs.

Feed and Speed Parameters

The options in the **F/S** tab are used to define parameters related feed and speed for machining; refer to Figure-3. The options in this tab are discussed next.

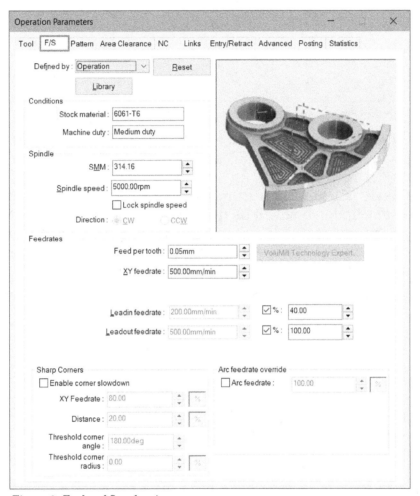

Figure-3. Feed and Speed options

- Select desired option from the **Defined by** drop-down to define source for feed and speed information. Select the **Operation** option from the drop-down if you want to use parameters specified in **Operation Parameters** dialog box in current tab.

Library Options

- Select the **Library** option from the **Defined by** drop-down to use parameters specified in material library. On selecting the **Library** option, the **Library** button below the drop-down becomes active. Click on the **Library** button to access **Micro Estimating Material Library** dialog box; refer to Figure-4.

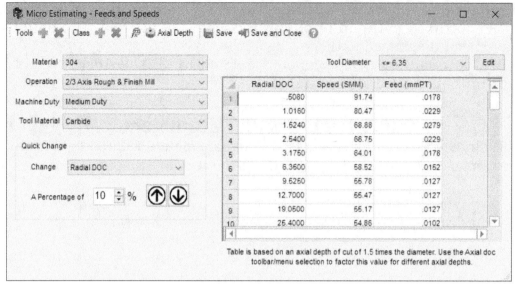

Figure-4. Micro Estimating Material Library dialog box

- Select desired workpiece material, operation, machine duty (class), and tool material for which you want to specify speed and feed rates. Specify the parameters in the table as desired.
- Click on the **Save Close** option from the toolbar in the dialog box to exit. The **Material Library** dialog box will be displayed; refer to Figure-5. You can specify cutting data for more materials by using the **Add Material** button in dialog box. Click on the **Close** button from **Material Library** dialog box to exit. The **Operation Parameters** dialog box will be displayed again.

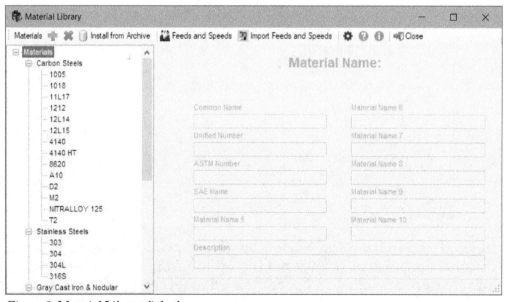

Figure-5. Material Library dialog box

- Select the **Tool** option from the **Defined by** drop-down to use speed and feed parameters specified in the cutting tool data. In our case, we will use **Operation** option of the **Defined by** drop-down.
- The options of the **Stock material** and **Machine duty** edit boxes cannot be changed in this dialog box. You can modify these parameters in the **Machine** dialog box and **Stock Manager**.
- Specify desired surface feet per minute or surface meter per minute value in **SFM** or **SMM** edit boxes, respectively. It is the product of diameter of tool and its

rotational speed in RPM.

- Select the **Lock spindle speed** check box if you want to keep RPM fixed even when diameter of tool has changed.

The options in the **Feedrates** area are used to define parameters at which cutting will occur during machining.

- Specify desired value of tool movement per tooth in cutting direction in the **Feed per tooth** edit box. So, when tool will move while cutting then tool will advance by specified value per tooth.
- Specify desired value in **XY feedrate** edit box to define tool movement rate in XY plane while cutting.
- Specify desired value in **Leadin feedrate** edit box to define the speed at which tool will enter into cutting pass.
- Specify desired value in **Leadout feedrate** edit box to define the speed at which tool will exit after cutting material.
- Select the **%** check box for Leadin feedrate or Leadout feedrate if you want to specify feedrate in percentage of XY feedrate value specified.
- Select the **Enable corner slowdown** check box if you want to reduce the tool movement speed around sharp corners. The options to define feed values will be displayed; refer to Figure-6.

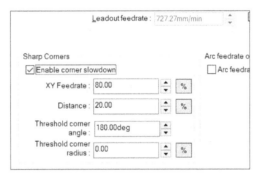

Figure-6. Feed options for sharp corners

- Specify desired value of **XY Feedrate** in percentage of normal XY Feedrate if **%** button is selected. De-select the **%** button if you want to specify the value of feedrate.
- Specify desired value of distance at which tool will slowdown before cutting around sharp corners in the **Distance** edit box of **Sharp Corners** area.
- Specify desired angle value in **Threshold corner angle** edit box to define the range up to which intersection of faces will be considered as sharp corner. If the angle between two intersecting faces is more than specified value then it will not be machined using sharp corner strategy.
- Specify desired value of radius below which the corner will be considered as sharp corner in the **Threshold corner radius** edit box. By default, **0** is specified in this edit box.
- Select the **Arc feedrate** check box from the **Arc feedrate override** area to define the speed at which tool will move around arcs.

Pattern Parameters

The options in the **Pattern** tab are available for **Area Clearance** operation; refer to Figure-7. The options in this tab are discussed next.

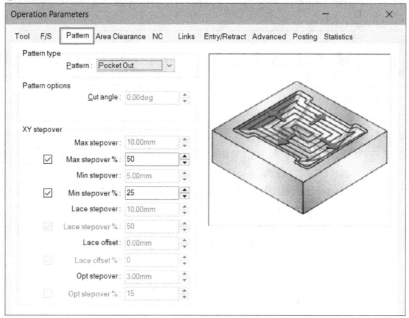

Figure-7. Pattern tab options

- Select the **Pocket Out** option from the **Pattern** drop-down if you want to create a toolpath which offset outward from center. Note that depending on shape of pocket there can be multiple different section of toolpath offsetting outward from their respective centers. Select the **Pocket In - Core** option from the **Pattern** drop-down if you want to create a toolpath which offset toward identified core/ boss feature from outer boundary. Select the **Lace** option from the **Pattern** drop-down if you want to create parallel toolpaths along a line at user defined angle; refer to Figure-8.

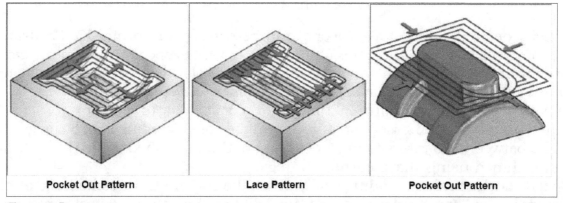

| Pocket Out Pattern | Lace Pattern | Pocket Out Pattern |

Figure-8. Pattern types

- The value in **Max stepover** edit box cannot be changed manually because this value is dependent on diameter of cutting tool. The maximum stepover allowed is equal to diameter of cutting tool.
- Specify desired values in **Max stepover %** and **Min stepover %** edit boxes to define maximum and minimum difference between two consecutive offset toolpaths. These edit boxes are available for pocket out patterns.
- For Lace pattern: Specify desired value of distance between two parallel lines of

toolpath in the **Lace stepover %** edit box. Specify desired value of distance of toolpath from the boundary of part in the **Lace offset %** edit box. Specify desired value in **Cut angle** edit box to define tool direction for lace cutting.

• For Pocket In - Core pattern: Specify desired values in **Max stepover %** and **Min stepover %** edit boxes to define maximum and minimum difference between two consecutive offset toolpaths.

These parameters are shown in Figure-9.

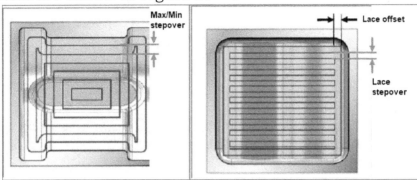

Figure-9. Pattern parameters

Area Clearance Tab

The options in the **Area Clearance** tab are available for Area Clearance operations. The options in this tab are used to define parameters related to area clearance toolpaths; refer to Figure-10. These options are discussed next.

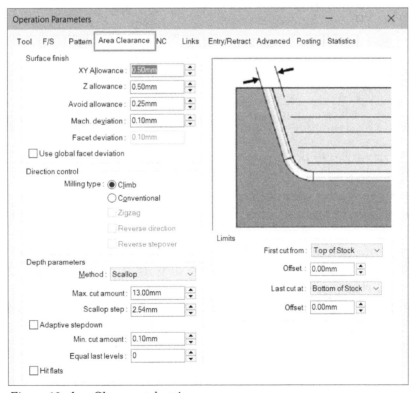

Figure-10. Area Clearance tab options

• Specify desired value in **XY Allowance** edit box to define allowable distance from the part wall where tool can start cutting in XY plane.

• Specify desired value in **Z allowance** edit box to define allowable distance from the bottom of part up to which tool will perform cutting while going downward.

• Specify desired value in **Avoid allowance** edit box to define distance allowable from walls of avoid section up to which tool can perform cutting.

- Specify desired value in `Mach. deviation` edit box to define the maximum deviation from curved faces of part; refer to Figure-11.

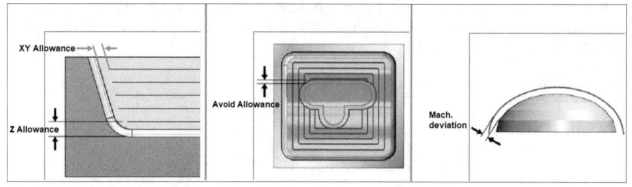

Figure-11. Surface finish parameters

- Select the `Use global facet deviation` check box to use facet deviation value defined in `Mill Features` tab of the `Options` dialog box displayed on clicking the `SOLIDWORKS CAM Options` tool.

- Select desired radio button for `Milling type` from the `Direction control` area. There are two ways for defining direction control; Climb and Conventional; refer to Figure-12.

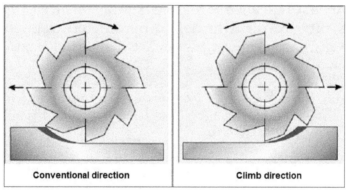

Figure-12. Cutting directions

- Select the `Zigzag` check box to perform cutting in both forward and backward directions alternatively while cutting in XY plane.

- Select the `Reverse direction` check box to reverse cutting direction on toolpath.

- Select the `Reverse stepover` check box to reverse the direction of stepover for lace pattern.

- Select desired option from the `Method` drop-down in the `Depth parameters` area to define motion while cutting tool moves downward. Select the `Constant` option from the `Method` drop-down to move tool downward by specified cut amount in each cutting pass. Select the `Scallop` option from the `Method` drop-down if you want to use cut amount based on scallop steps. Scallop is the material left on the theoretical surface between two successive tool passes; refer to Figure-13.

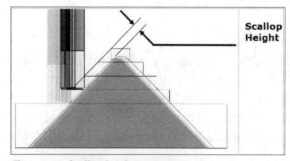

Figure-13. Scallop height parameter

- Specify desired value in **Max. cut amount** and **Scallop step/Cut amount** edit boxes. The **Scallop** step edit box is displayed when **Scallop** option is selected in the **Method** drop-down.
- On selecting the **Adaptive stepdown** check box, the tool will make first cut at maximum depth and then based on left material, the depth will be reduced automatically for better finish.
- Specify the minimum amount of cut that can be made during adaptive stepdown in the **Min. cut amount** edit box.
- If you want to perform multiple cuts of equal depth when reaching the minimum Z height in order to avoid a very shallow final level, specify the respective number of levels to be considered equal in **Equal last levels** edit box.
- Select the **Hit flats** check box to automatically add a cutting pass for machining at depth specified in **Z allowance** edit box.
- Specify desired parameters in the **Limits** area to define upper and lower limits for cutting operation.
- Select desired option from the **First cut from** drop-down to define reference to be used for making first/top cut.
- Specify desired value in **Offset** edit box to define distance from selected reference at which first cut will be made.
- Select desired option from the **Last cut at** drop-down to select reference last cut.
- Specify desired value in **Offset** edit box as discussed earlier for bottom cut.

NC Tab

The options in the **NC** tab are used to define position of retract plane and home point; refer to Figure-14. These options are discussed next.

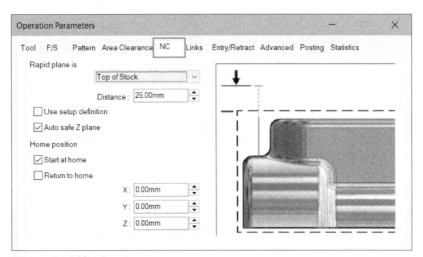

Figure-14. NC tab options

- Select desired option from the **Rapid plane is** drop-down to define reference for rapid plane. Rapid plane is the plane from where tool begins cutting and after completion of cutting operation, the tool retracts back to it.
- Specify desired value in **Distance** edit box to define distance from selected rapid plane reference.
- Select the **Use setup definition** check box if you want to use machine setup for defining rapid plane.

- Select the **Auto safe Z plane** check box if you want SolidWorks CAM to automatically find location of safe Z plane. If this safe Z value is greater than the Z position based on the Rapid plane method and Distance value, the safe Z value is used to define the Rapid plane.
- Select the **Start at home** check box to specify coordinates for tool from where tool will start cutting. After selecting this check box, specify X, Y, and Z coordinates of home position in respective edit boxes.
- Select the **Return to home** check box to move cutting tool back to home position after cutting.

Links Tab

The options in the **Links** tab are used to define parameters related to section linking toolpath passes; refer to Figure-15. The options in this tab are discussed next.

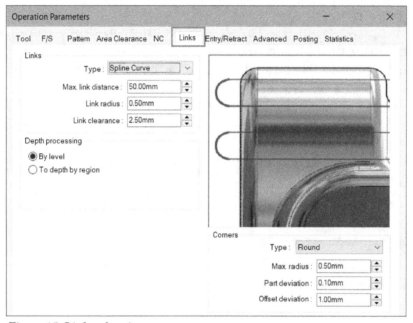

Figure-15. Links tab options

- Select desired option from the **Type** drop-down to define shape of link between two consecutive toolpath passes. Select the **Spline Curve** option from the drop-down to create spline link or select the **Straight Line** option from the drop-down to create straight line link between toolpath passes.
- Specify desired parameters in the **Max. link distance**, **Link radius**, and **Link clearance** edit boxes to define shape and size of link.
- Select desired radio button from the **Depth processing** area to define depth upto which machining will be performed. When the **By level** option is selected, roughing is completed for all areas of the model at a given Z depth before machining the next Z depth. When the **To depth by region** option is selected, roughing is completed to the bottom of each area in the model before machining another area. Selecting this option can decrease machining time.
- Select desired option from the **Type** drop-down in the **Corners** area of the dialog box to define whether corners are sharp or round. If **Round** option is selected in the **Type** drop-down then you can specify parameters like maximum round radius, deviation from part, and offset deviation in respective edit boxes of the **Corners** area.

Entry/Retract Tab

The options in the **Entry/Retract** tab are used to define method and parameters for entry and retraction of tool while cutting; refer to Figure-16. The options in this tab are discussed next.

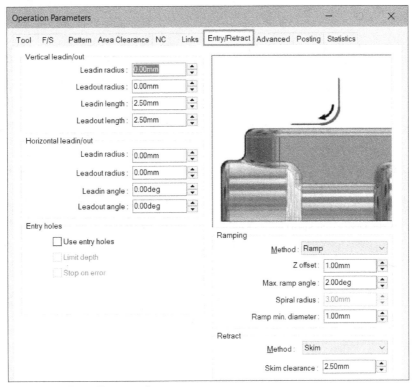

Figure-16. EntryRetract tab

For Vertical leadin/out:

- Specify desired value of radius in the **Leadin radius** edit box to define radius of arc when tool approaches the workpiece while cutting.
- Specify desired value of radius in the **Leadout radius** edit box to define radius of arc when tool exits the workpiece after cutting.
- Specify desired values of length for leadin and leadout in **Leadin length** and **Leadout length** edit boxes, respectively.
- You can similarly specify parameters related to leadin/leadout for horizontal direction in the **Horizontal leadin/out** area.
- Select the **Use entry holes** check box if you have created entry holes for machining and want to use them for starting machining operation. On selecting this check box, a selection button will be displayed adjacent to the check box. Click on the selection button, the **Select Point** dialog box will be displayed; refer to Figure-17. Select the points on the model to be used for holes and specify the offset values in **Offset** area as desired. Click on the **OK** button from the dialog box to define points.

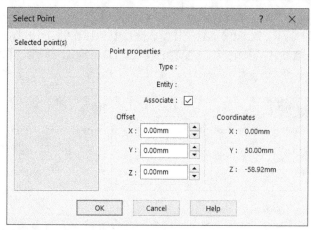

Figure-17. Select Point dialog box

- Select the **Limit depth** check box to limit the depth of machining by Z value specified for entry holes.
- Select the **Stop on error** check box to validate whether adequate number of entry holes have been specified. Note that when this check box is selected then none of the toolpaths will be generated if adequate number of entry holes have not been created.

Ramping Area

The options in the **Ramping** area are used to define how cutting tool will come downward while cutting.

- Select desired option from the **Method** drop-down to define cutting tool motion. Select the **Spiral** option from the drop-down if you want the cutting tool to spiral down during machining while entering the workpiece. Select the **Ramp** option from the drop-down if you want the cutting tool to come down in ramp motion while entering the workpiece. Select the **Plunge** option from the drop-down if you want the cutting tool to directly move downward into workpiece when starting the machining operation; refer to Figure-18.

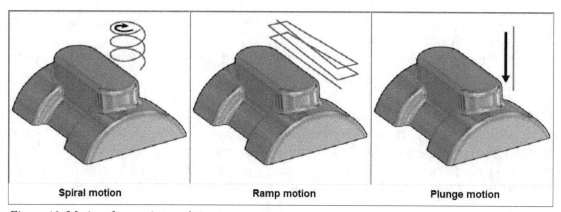

Figure-18. Motions for entering workpiece

- Set desired parameters in the **Ramping** area for entering motion of tool.
- Select desired option from the **Method** drop-down in the **Retract** area of the dialog box. Select the **Skim** option from the drop-down if you want the tool to rapid vertically to minimum Z clearance height. The minimum clearance height for skim motion can be specified in the **Skim clearance** edit box of **Retract** area in

the dialog box. Select the **Full** option from **Method** drop-down in **Retract** area if you want the cutting tool to retract back to rapid plane.

Geometry Tab

The options in the **Geometry** tab are used to define faces to be cut, faces to be avoided, containment area, and user defined avoid area; refer to Figure-19. Various options of this tab are discussed next.

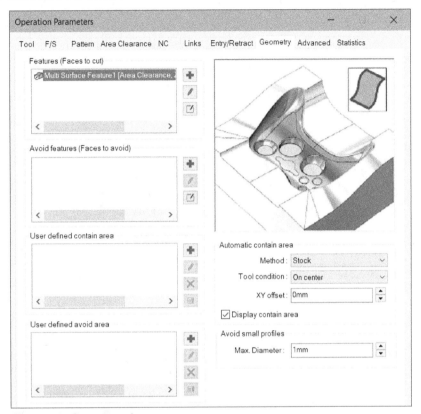

Figure-19. Geometry tab

- Click on the **Create** button from **Features (Faces to cut)** area to create additional faces/features to be machined by the toolpath. On clicking this button, the **Multi Surface Feature PropertyManager** will be displayed as discussed earlier.
- To edit any feature earlier selected for cutting, select the feature from **Features (Faces to cut)** area and click on the **Edit** button. The **PropertyManager** to edit feature will be displayed as discussed earlier.
- Click on the **Advanced** button from the **Features (Faces to cut)** area to define the features to be included in the machining. The **Select features** dialog box will be displayed; refer to Figure-20. Select all the features to be included in the list and click on the **OK** button.
- Similarly, you can define rest of the features required for machining from other areas of the dialog box.
- Select desired option from the **Method** drop-down in **Automatic contain area** section to define how containment area of toolpath will be defined. Select the **Stock** option to use boundary of stock as containment area. Select the Bounding Box option from the drop-down to use bounding box for the part as containment area. Select the **All Silhouettes** option from the drop-down to use boundary edges of the part as containment area. Select the **Outer Silhouettes** option from the drop-down to ignore inner sections of part when creating boundary. The effect of these options is shown in Figure-21.

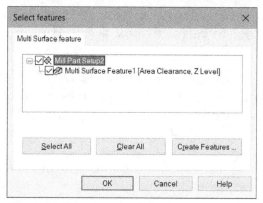

Figure-20. Select features dialog box

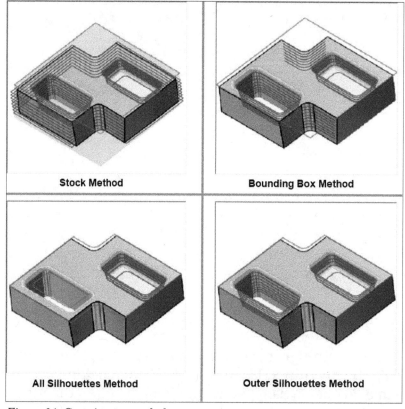

Figure-21. Containment methods

- Select the **On center** option from the **Tool condition** drop-down to move center of tool over toolpath curves while cutting. Select the **Upto** option from the **Tool condition** drop-down to move the cutting tool inside the boundary in such a way that edge of tool moves over the toolpath. Select the **Past** option from the drop-down to move the cutting tool outside the boundary in such a way that edge of tool moves over the toolpath. The tool positions for these options are shown in Figure-22.
- Specify desired value in the **XY offset** edit box for offsetting the containment boundary by specified value.
- Specify desired value in the **Max. Diameter** edit box of **Avoid small profiles** area to define maximum diameter of small features which will not be included in machining toolpath.

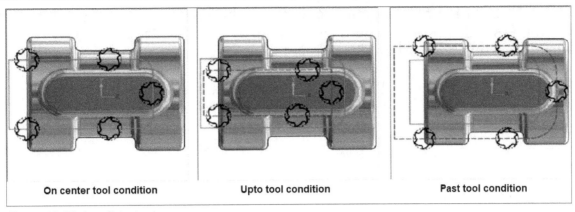

Figure-22. Tool condition options

Advanced Tab

The options in the **Advanced** tab are used to define containment area, arc deviation, holder avoidance limit, and toolpath mirroring parameters; refer to Figure-23. Various options in this tab are discussed next.

Figure-23. Advanced tab

- Select the **Line moves only** check box if your machine do not support G2/G3 codes which are used for circular interpolation. All the arcs will be converted to chain of small lines.
- Specify desired values for deviation from arc toolpath in the **Deviation** and **Chordal deviation** edit boxes in the **Arc fitting** area.
- Select the **Enable** check box from the **Holder avoidance** area to avoid contact of tool holder from workpiece by specified clearance value. The value for clearance can be specified in **Clearance** edit box after selecting the check box.
- Select the **Calculate min tool protrusion** check box to automatically calculate the length of tool for tool holder clearance. After selecting this check box, specify desired value in the **Clearance** edit box. Now, click on the **Preview** button from the dialog box to automatically calculate the minimum length required. Once

preview simulation run is complete, click on the tabs of the dialog box to display parameters. The updated length will be displayed in the **Calculated min** edit box.

- Select the **Mirror toolpath** check box to mirror current toolpath with respect to selected reference. To select a reference, click on the **Mirror Entity** button from the **Mirror** area of the dialog box. The **Select Axis** dialog box will be displayed; refer to Figure-24. Select desired edge/axis and click on the **OK** button.

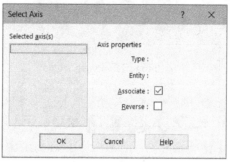

Figure-24. Select Axis dialog box

- Select the **Keep original** check box to keep original toolpaths after mirroring. If you want to delete the original after copying then clear the **Keep original** check box.
- Set desired values in **X offset** and **Y offset** edit boxes to offset the mirror copy of toolpath.
- Select the **Use Setup Definition** check box to use part setup parameters for mirroring the toolpaths.

Statistics Tab

The options in this tab are used to display various toolpath related statistics like number of line segments and arc segments, length travelled at feed speed, length travelled at rapid speed, non-cutting time, total machining time of toolpath, and so on.

- If you want to save current operation parameters in the technical database then click on the **Save As Defaults** button from the **TechDB** area of the Statistics tab in the dialog box. The **Save As Defaults** dialog box will be displayed; refer to Figure-25. Specify desired name and description of operation plan in the **Name** and **Description** edit boxes. Select the **Set As Default** check box to make newly created operation plan as default for area clearance. Click on the **OK** button from the dialog box to save the operation plan.

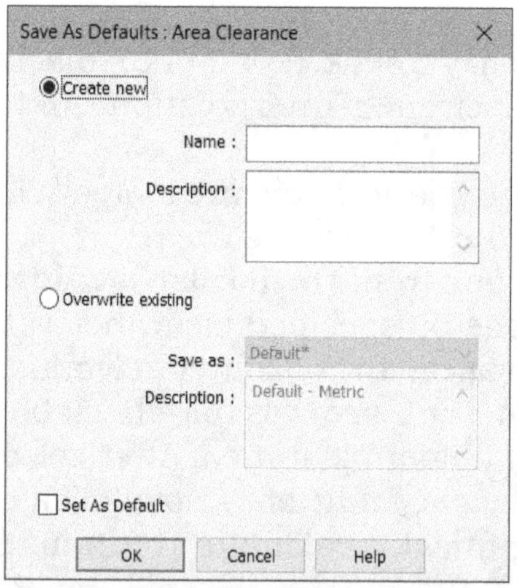

Figure-25. Save As Defaults dialog box

- If you want to load default saved operation plans from the technical database then click on the **Load Defaults** button. The **Load Defaults** dialog box will be displayed with default operation plans for current operation; refer to Figure-26. Select desired option from the **Defaults Name** drop-down and click on the **OK** button.

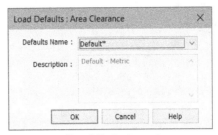

Figure-26. Load Defaults option

- After setting desired parameters, click on the **OK** button from the **Operation Parameters** dialog box.

There are three tools to create mill operations manually which are **2.5 Axis Mill Operations**, **Hole Machining Operations**, and **3 Axis Mill Operations**. These tools are discussed next.

CREATING 2.5 AXIS MILL OPERATIONS MANUALLY

The tools in **2.5 Axis Mill Operations** cascading menu are used to create different type of 2.5 axis mill operations. The procedures to use these tool is given next.

- Click on the **Rough Mill** tool or any other operation creation tool from **Tools > SOLIDWORKS CAM > New > 2.5 Axis Mill Operations** cascading menu from the menu bar; refer to Figure-27. The **New Operation PropertyManager** will be displayed; refer to Figure-28.

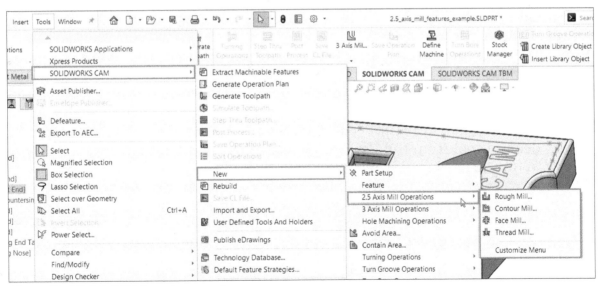

Figure-27. 2.5 Axis Mill Operations cascading menu

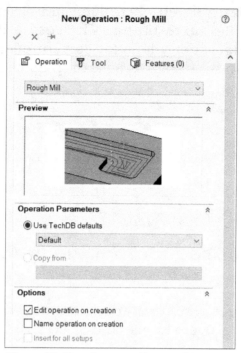

Figure-28. New Operation PropertyManager

- Select desired operation type from the drop-down at the top in **Operation** tab of the **PropertyManager**.

Creating Face Mill Operation

- Select the **Face Mill** option from the top drop-down in the **Operation** tab.
- Select the **Use TechDB defaults** radio button from the **Operation Parameters** rollout to use default parameters saved in database for operation. Select desired option from the drop-down displayed in selecting the radio button.
- If there are multiple operations saved in current project then you can select another operation for copying parameters by selecting the **Copy from** radio button from the **Operation Parameters** rollout.
- Select the **Edit operation on creation** check box from the **Options** rollout to edit the operation after creating it.
- Select the **Name operation on creation** check box to specify name of operation after creating it.
- Click on the **Tool** tab from the **PropertyManager** to select the tool; refer to Figure-29.
- Select desired tool from the list box to use it for facing operation. It is generally better to use large diameter face milling tool for facing operation.
- Click on the **Features** tab from the **PropertyManager** to select the feature for facing operation; refer to Figure-30. If face feature has not been created earlier then click on the **Create Features** button from the **Features for the Face Mill** rollout and create the feature as discussed in previous chapter.

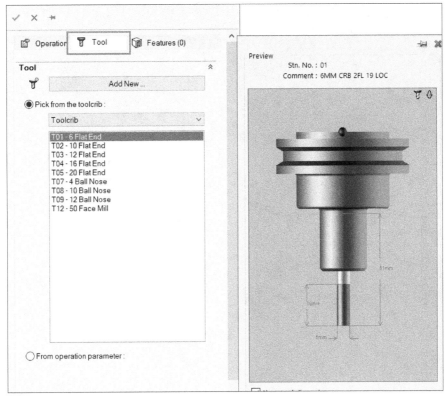

Figure-29. Tool tab

- After selecting feature from the **Pick from the available** area of the **PropertyManager**, click on the **OK** button from the **PropertyManager**. If you have selected the **Name operation on creation** check box, then the **Properties** dialog box will be displayed; refer to Figure-31.

Figure-30. Features tab

Figure-31. Properties dialog box

- Specify desired name and description for the operation and click on the **OK** button. If the **Edit operation on creation** check box is selected in the **New Operation PropertyManager** then the **Operation Parameters** dialog box will be displayed with facing options; refer to Figure-32.

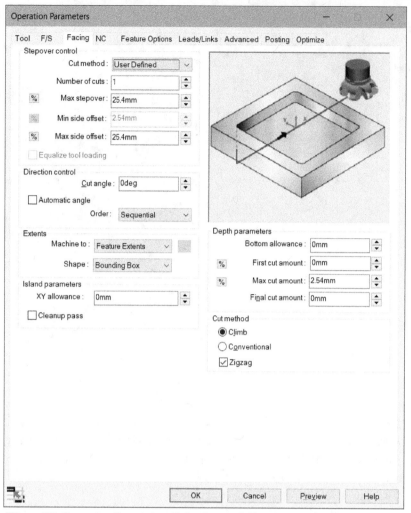

Figure–32. Operation Parameters dialog box with Facing options

Most of the options are same as discussed for area clearance operation. The other options are discussed next.

Feed and Speed Parameters

*   You can set system calculated feedrate for machining arcs by selecting the System calculated feedrates check box from the **Arc feedrate override** area of **F/S** tab in the dialog box. On selecting this check box, specify the minimum and maximum level of feedrates for arcs in the **Min. feedrate** and **Max feedrate** edit boxes, respectively.

Facing Parameters

*   Select desired option from the **Cut method** drop-down of **Facing** tab to define how facing cuts will be made. Select the **User Defined** option from the drop-down if you want to specify number of cutting passes for facing. Select the **Automatic** option from the **Cut method** drop-down if you want to define number of cutting passes automatically based on size of workpiece.
*   Specify desired value of maximum stepover in the **Max stepover** edit box. Note that you should not specify maximum stepover value more than diameter of facing tool.
*   Specify desired value in the **Min side offset** edit box to define minimum distance from the part walls upto which facing tool can offset while machining.
*   Specify desired value in the **Max side offset** edit box to define maximum distance from the part walls upto which facing tool can offset while machining.

- Select the **Equalize tool loading** check box from the **Stepover control** area to modify the toolpaths in such a way that facing tool is equally loaded.
- Specify desired angle value in **Cut angle** edit box to reorient the toolpath at specified angle. Select the **Automatic angle** check box to let SolidWorks CAM automatically decide the cut angle from the **Direction control** area.
- Select the **Sequential** option from the **Order** drop-down if you want to perform cutting in sequence. Select the **Out To In** option from the drop-down if you want to perform cutting in different sections starting from outside and then moving inward.
- Select the **Feature Extents** option from the **Machine to** drop-down of **Extents** area to use extents of face feature for defining machining boundary. Select the **To WIP** option from the **Machine to** drop-down to use current stock boundaries after previous operations have been performed. After selecting the **To WIP** option from the drop-down, click on the **Browse** button from **Extents** area. The **Operations for WIP** dialog box will be displayed; refer to Figure-33. Select desired part from the dialog box and click on the **OK** button.

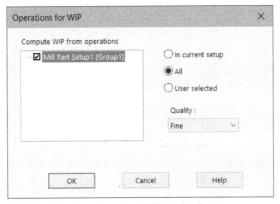

Figure-33. Operations for WIP dialog box

- Select the **Silhouette** or **Bounding Box** option from the **Shape** drop-down to define the shape of boundary to be used for facing operation.
- Specify desired offset value in the **XY allowance** edit box if there are islands on the model face and you want to specify offset distance for island boundary.
- Select the **Cleanup pass** check box to create one more toolpath for finishing the island edges.
- Specify desired parameters in the **Depth parameters** area to define depth of facing cuts. Specify desired value in **Bottom allowance** edit box to define the tolerance from base of part up to which facing need to be performed. Specify desired value in the **First cut amount** edit box to define depth of first facing cut. Specify desired value in the **Max cut amount** edit box to define depth of maximum cut that can be performed during facing operation. Specify desired value in the **Final cut amount** edit box to define finishing move of facing operation.
- Set desired parameters in the **Cut method** area to define cutting direction as discussed earlier.

Feature Options Parameters

- Select desired option from the **Feature list** area of **Feature Options** tab to define feature to used for defining parameters. After selecting feature, click on the **Parameters** button from the dialog box. The **Face Feature Parameters** dialog box will be displayed; refer to Figure-34.

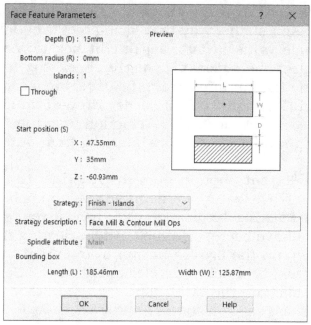

Figure-34. Face Feature Parameters dialog box

- If you want to modify the strategy of facing operation, select desired option from the **Strategy** drop-down.
- Select the **Through** check box if you want to perform facing through the workpiece. After setting parameters, click on the **OK** button from the dialog box.
- Click on the dimension button [icon] next to `Machining depth` field and specify the depth value in the edit box.

Optimize Parameters

- Click on the **Optimize** tab to display optimization parameter. Select desired option from the **Method** drop-down of **Optimization** area to define optimization method. Select the **None** option if you do not want to optimize order of toolpaths for faster material removal. In such case, the order in which operations are placed in the **SOLIDWORKS CAM Feature Tree** will be used. Select the **Grid** option from the **Method** drop-down if you want to specify grid points for machining. Select the **Shortest Path** option from the **Method** drop-down if you want to use shortest paths for performing machining. Select the **Inside to Out** option from the drop-down if you want the tool to move inward while cutting from outer edge. Select the **Outside to In** option from the drop-down if you want the tool to move outward from center.
- Select desired radio button from the **Start point** area of the dialog box to define start point of toolpath. Select the **Corner** radio button and select desired option from next drop-down to define starting point. Select the **Entity select** radio button and click in the next selection box to select the start point from model.
- Select the **Last closest** check box to use last point of previous operation as start point of current operation.
- If you have selected **Grid** option from the **Method** drop-down then options in **Grid parameters** area will be active. Select desired option from the **Direction** drop-down to define direction of grid. Select the **Zig** or **Zigzag** option from the **Pattern** drop-down to how tool will progress while cutting. Specify desired value in **Band width** edit box to define gap between two consecutive instances of grid points.
- The parameters in **Toolpath analysis** and **Estimated machining time** area are updated automatically when you preview the toolpath.
- After setting desired parameters, click on the **OK** button to create toolpath operation.

Creating Rough Mill Operation

The rough mill operation is used to machine various mill features like slots and pockets. The procedure to create rough mill operation is given next.

* Select the **Rough Mill** option from the **Operation Type** drop-down in the **Operation** tab of **New Operation PropertyManager**.
* Select desired option from the drop-down in **Operation Parameters** rollout after selecting the **Use TechDB defaults** radio button. Select the **Volumill** option from the drop-down to use volume mill strategy. If most of the features in model are boss features then you can select the **Core machining** option from drop-down to perform machining based on core milling strategy. Similarly, you can select **Cavity machining** option to perform machining in a mold cavity.
* Select the check boxes from the **Options** rollout as discussed earlier.
* Click on the **Tool** tab and select desired cutting tool.
* Click on the **Features** tab in the **PropertyManager** to select features to be used for rough machining. You can create features for rough mill by using the **Create Features** button as discussed earlier.
* After selecting features, click on the **OK** button from the **PropertyManager**. If check boxes are selected in the **Options** rollout then **Properties** dialog box will be displayed.
* Click on the **OK** button from the **Properties** dialog box. The **Operation Parameters** dialog box will be displayed for rough mill operation; refer to Figure-35.

Figure-35. Operation Parameters dialog box with Roughing options

* Select desired option from the **Pattern** drop-down to define pattern in which machining should be performed. The progress of tool with various patterns will be displayed; refer to Figure-36. Based on selected option in the drop-down, the parameters will be displayed in **Pocketing** area.

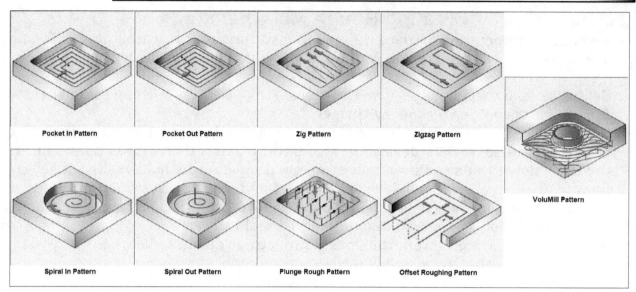

Figure-36. Pattern types for roughing

- If you have selected the **VoluMill** option then you will be asked to use VoluMill Technology Expert for changing cut parameters. Click on the **Yes** button from **SOLIDWORKS CAM Warning** dialog box displayed on selecting the **VoluMill** option. The **VoluMill Technology Expert** dialog box will be displayed; refer to Figure-37.

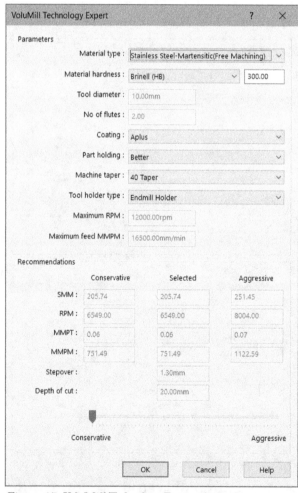

Figure-37. VoluMill Technology Expert dialog box

- Select desired material from the **Material type** drop-down to define material of workpiece.

- Select desired option from the **Material hardness** drop-down to define unit for hardness and specify the hardness of material in adjacent edit box.

- Select desired option from the **Coating** drop-down to define coating applied on cutting tool.

- Select desired option from the **Part holding** drop-down to define how well your workpiece is fixed on the table. If part is held better on table then faster feed can be applied for machining.

- Select desired option from the **Machine taper** drop-down to define spindle taper for installing related types of tool holders and arbors.

- Select desired option from the **Tool holder type** drop-down to define tool holder type of machine.

- Move the slider in **Depth of cut** area to define whether you want to specify depth of cut as conservative or aggressive. Note that based on specified parameters, the parameters related to machining will be displayed in the **Recommendations** area like SMM, RPM, MMPT, and so on.

- After setting desired parameters, click on the **OK** button from the dialog box. Note that you can open the **VoluMill Technology Expert** dialog box by clicking on the **VoluMill Technology Expert** button from the **F/S** tab of the dialog box.

- If you have selected option other than **VoluMill** from the **Pattern** drop-down then specify the related parameters in the **Pocketing** area as discussed earlier.

- Specify desired value of tolerance up to which machining will be performed around the side walls of pocket in the **Allowance** edit box in the **Side parameters** area of **Roughing** tab.

- Specify desired value of stepover in the **Stepover** edit box for side walls in the **Side parameters** area of the tab.

- Select the **Wedge machining** check box from the **Side parameters** area to machine wedges created by pocket in or pocket out pattern toolpaths with stepover more than 50% of tool diameter.

- Select the **Generate** check box from the **Rest machining** area of the dialog box to display and generate toolpath for areas which have not been machined by previous rough machining operation.

- Select desired option from the **Machine** drop-down in the **Rest machining** area to define which stocks are to be machined. Select the **No** option if you want to machine every area without restricting to any specific operation leftover. Select the **From WIP** option from the drop-down if you want to select an operation based on which the leftover workpiece material will be calculated. After selecting the **From WIP** option, click on the **Browse** button next to drop-down. The **Operations for WIP** dialog box will be displayed; refer to Figure-38. Set desired parameters in the dialog box and click on the **OK** button to select operation. Select the **Previous Leftover** option from the **Machine** drop-down if you want to machine the areas that have not been machined by first rough mill operation.

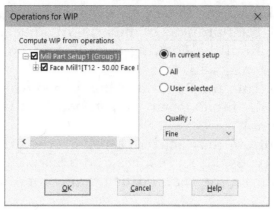

Figure-38. Operations for WIP dialog box

- Set the other parameters of **Roughing** tab as discussed earlier.
- If you want to specify cutter compensations then select the **On** option from the **CNC compensation** area of the **NC** tab in dialog box.
- The other options in the **Operation Parameters** dialog box are same as discussed earlier. Click on the **OK** button to create the operation.

Creating Contour Mill Operation

The contour milling is used to finish walls of boss and pocket features.

- Select the **Contour Mill** option from the **Operation Type** drop-down of **New Operation PropertyManager**.
- Select desired option from the **TechDB defaults** drop-down; refer to Figure-39 to define strategy for machining.

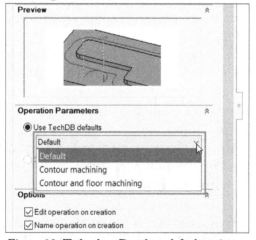

Figure-39. Technology Database default options

- Select desired cutting tool and features from the **Tool** and **Features** tab of the **PropertyManager**.
- Click on the **OK** button from the **PropertyManager**. The **Properties** dialog box will be displayed. Specify desired name and description of operation, and click on the **OK** button from the dialog box. The **Operation Parameters** dialog box will be displayed; refer to Figure-40.

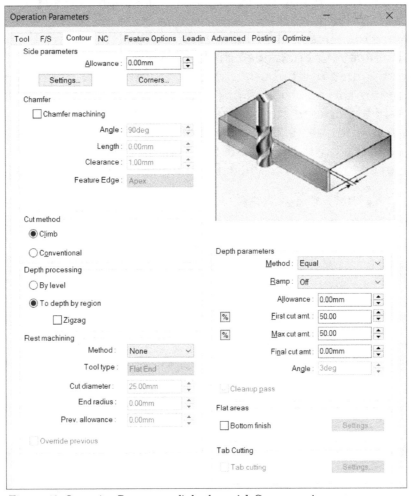

Figure-40. Operation Parameters dialog box with Contour options

- Specify desired value of tolerance in the **Allowance** edit box to define deviation around the walls.
- Click on the **Settings** button from the **Side parameters** area to define settings related to side parameters. The **Side Parameters** dialog box will be displayed; refer to Figure-41.

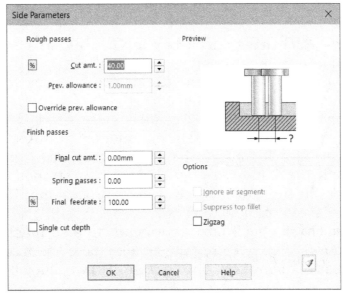

Figure-41. Side Parameters dialog box

- Specify desired value in **Cut amt.** edit box of **Rough passes** area to define the depth of cut around contours.
- Select the **Override prev. allowance** check box to override the value of wall tolerance in the **Prev. allowance** edit box.
- Specify desired values of final cut amount, number of final passes, and feedrate for final passes in the edit boxes of **Finish passes** area. Select the **Single cut depth** check box to perform single finish pass. Set the other parameters and click on the **OK** button from the **Side Parameters** dialog box.
- Click on the **Corners** button from the **Side parameters** area of the dialog box. The **Corner Parameters** dialog box will be displayed; refer to Figure-42.

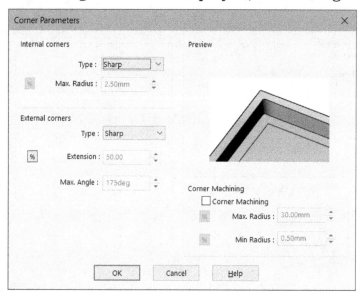

Figure-42. Corner Parameters dialog box

- Select desired option from the **Type** drop-down in **Internal corners** area of the dialog box to define whether corners will be sharp or round by specified radius.
- Similarly, you can set the external corners sharp or round by specifying the parameters in the **External corners** area.
- Select the **Corner Machining** check box from the **Corner Machining** area to specify maximum or minimum radius range.
- Click on the **OK** button from the **Corner Parameters** dialog box to apply corner cutting parameters.
- Select the **Chamfer Machining** check box in **Chamfer** area if you are machining chamfers on the contour lines of part. Specify the length of chamfer in the **Length** edit box. Specify desired value in the **Clearance** edit box to define depth from chamfer edge up to which cutting tool can go.
- The options in the **Feature Edge** drop-down are used to define whether the geometry in the feature is the apex of the chamfer or the outer edge of the chamfer.
- If you are chamfering a 3D curve then you can specify the **Avoid Allowance** parameter to specify the distance by which the tool is supposed to avoid that edge.
- Set the parameters in the Cut method, Depth processing, Rest machining, and Depth parameters areas of the dialog box.
- Select the **Bottom finish** check box if you want to also machine the bottom flat faces of the contours. After selecting the check box, click on the **Settings** button from the **Flat areas** of dialog box. The **Bottom Finish** dialog box will be displayed; refer to Figure-43.

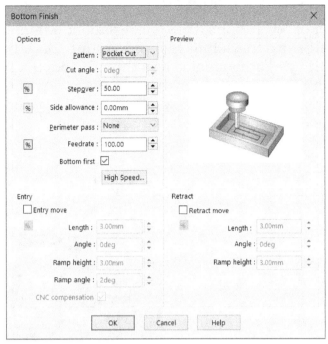

Figure-43. Bottom Finish dialog box

- The parameters in this dialog box have been discussed earlier. Set desired parameters for bottom finish and click on the **OK** button.
- Select the **Arc fit** check box from the **Spline output** area of the **Advanced** tab in the dialog box if you want to use arcs for cutting tool movements rather than using spline points.
- Set the other parameters as desired and click on the **OK** button to create contour machining operation.

Creating Thread Mill Operation

The thread mill operation is used to create internal and external threads. The procedure to create this operation is given next.

- Select the **Thread Mill** option from the **Operation Type** drop-down in the **New Operation PropertyManager**.
- Set the other parameters like cutting tool and features from the **PropertyManager**, and click on the **OK** button. The **Operation Parameters** dialog box will be displayed with parameters related to threading; refer to Figure-44.
- Based on the feature selected in **Feature list** drop-down of **Feature Options** tab of the dialog box, the thread parameters are displayed in the **Thread parameters** area of the **Thread Parameters** tab in the dialog box.
- Specify desired major diameter, minor diameter, pitch, taper angle, and angular resolution values in **Thread parameters** area of the dialog box.
- Select desired radio button from **Direction** area to define threading direction.
- If you have selected a multi-point threading tool, then options of the **Multi-point threading** area will be active. Select the **Single Point** option from the **Method** drop-down if you want to perform single point threading. The number of revolutions will be automatically calculated based on machining depth divided by pitch. Select the **User Defined** option from the **Method** drop-down if you want to specify number of helical revolutions manually. Select the **Automatic** option from the **Method** drop-down if you want SOLIDWORKS CAM to automatically calculate the number of revolutions.

- If you have selected the **User Defined** option from the **Method** drop-down then specify the number of revolutions, thread overlap, and minimum number of extra threads to be run before threading tool moves out. Note that thread overlap is required when cutting length of tool is less than the depth of thread and you need more than one run of cutting tool for threading.
- The other parameters in the dialog box have been discussed earlier. Click on the **OK** button from the dialog box to create the operation.

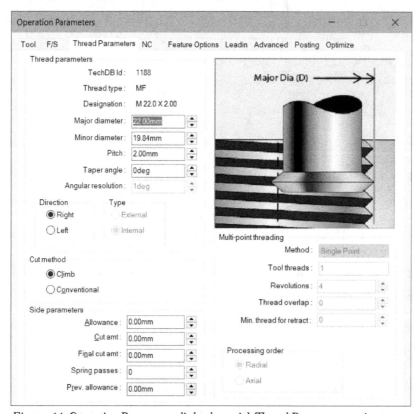

Figure-44. Operation Parameters dialog box with Thread Parameters options

CREATING HOLE MACHINING OPERATIONS

The tools in **Tools > SOLIDWORKS CAM > New > Hole Machining Operations** cascading menu are used to create holes by various methods like drilling, reaming, boring, and so on; refer to Figure-45. You can also activate **Hole Machining Operations** by selecting respective tools from **Hole Machining Operations** cascading menu in right-click shortcut menu when you right-click on **Mill Part Setup** node in **SOLIDWORKS CAM OPERATION TREE**; refer to Figure-46. The procedure to use these tools is discussed next.

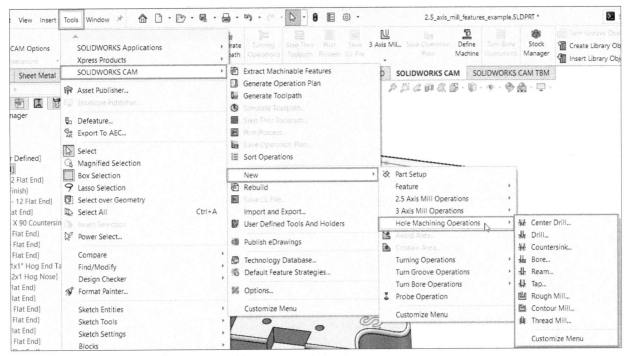

Figure-45. Hole Machining Operations cascading menu

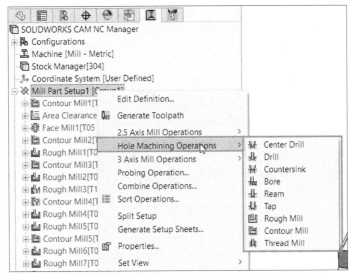

Figure-46. Hole Machining Operations shortcut menu

- Click on the **Center Drill** tool from the **Hole Machining Operations** cascading menu in **Tools > SOLIDWORKS CAM > New** menu. The **New Operation PropertyManager** will be displayed; refer to Figure-47.

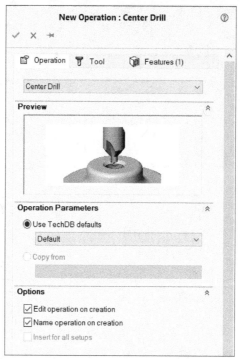

*Figure–47. New Operation PropertyManager
for Drill operations*

Creating Central Drill Operations

- Select the **Center Drill** option from the **Operation Type** drop-down in the **PropertyManager**.
- Select desired strategy for central drilling from the **TechDB defaults** drop-down in the **Operation Parameters** rollout; refer to Figure-48. You can create different type of strategies using the Technology Database options. Select the **Spot Drilling** option from drop-down to machine holes with depth less than cutting length of drill. Select the **Pecking** option to drill holes which have more depth then cutting length of drill bit and tool must retract by small amount between cutting passes.

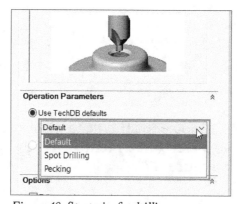

Figure–48. Strategies for drilling

- Select desired cutting tool and feature for drilling from the **PropertyManager** and click on the **OK** button. The **Operation Parameters** dialog box will be displayed; refer to Figure-49.

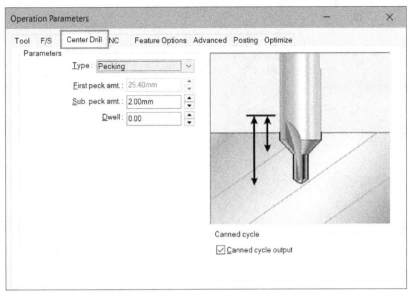

Figure-49. Operation Parameters dialog box with Center Drill options

- Select desired option from the **Type** drop-down to define drill operation type. Select the **Drilling** option from the drop-down if you want to drill hole up to specified depth. Select the **Spot Drilling** option from the drop-down if you want to drill a guide hole for other drilling operations. Select the **Pecking** option from the drop-down if you want to perform drilling in multiple successive small cuts so that chips get extracted during drilling. Select the **Variable Pecking** option from the drop-down if you want to perform drilling in varying depth sets. Select the **High Speed Pecking** option from the drop-down if you want to perform drilling in multiple high speed pecks.
- Specify the value of first drilling cut depth in the **First peck amt.** edit box.
- Specify desired value of depth in the **Sub. peck amt.** edit box to define value of depth for subsequent drilling pecks.
- Select the **Canned cycle output** check box if you want to output canned cycle NC codes like G73, G74, and so on.
- Specify the upper and lower limits of holes in the **Advanced** tab of dialog box.
- Set the other parameters as discussed earlier and click on the **OK** button to create the toolpath.

Creating Gun Drilling Operation

- Select the **Drill** option from the **Operation Type** drop-down in the **New Operation PropertyManager**. Set the other parameters in the **PropertyManager** as discussed earlier and click on the **OK** button. The **Operation Parameters** dialog box will be displayed with drill hole parameters; refer to Figure-50.

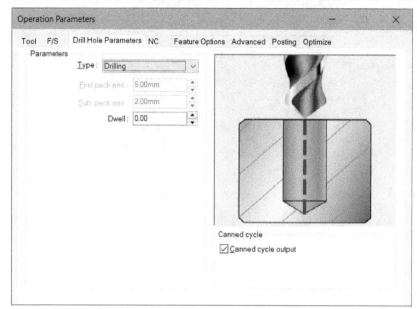

Figure-50. Operation Parameters dialog box with Drill Hole options

- Select the **Gun Drilling** option from the **Type** drop-down if you want to drill deep holes which have high hole depth to diameter ratio. Gun drilling can be performed for holes with diameter from 1 mm to 75 mm depending on availability of special equipment.
- Specify desired value in **Dwell** edit box to define time in seconds for which cutting tool will remain engaged before retracting.
- Specify desired value in the **Pilot entry/retract RPM** edit box to define the reduced RPM to be used for all pilot hole entry and retract moves, as well as all positioning moves that occur outside the pilot hole. Slow RPM is needed so that Gun drill can be manually guided into the pilot hole.
- Specify desired value in the **Pilot feed-in distance** edit box to define depth in pilot hole up to which cutting tool will feed in.
- Specify desired value of feed rate for entering pilot hole in the **Pilot feed-in feedrate** edit box.
- Select the **Pilot feed-in feedrate %** check box if you want to specify feedrate in percentage of feedrate specified in the **F/S** tab of the dialog box.
- Similarly, specify the other parameters in the dialog box and click on the **OK** button to create the operation.

Creating Countersink Holes

- Select the **Countersink** option from the **Operation Type** drop-down in **PropertyManager** and specify other parameters like cutting tool and features (holes) as discussed earlier.
- Click on the **OK** button from the **PropertyManager**. The **Operation Parameters** dialog box will be displayed with options related to drilling; refer to Figure-51.
- Set desired parameters in the dialog box as discussed earlier and click on the **OK** button to create the operations.

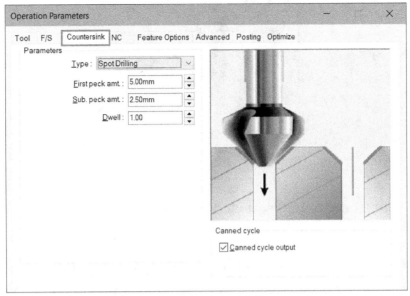

Figure-51. Operation Parameters dialog box with Countersink options

Creating Bore Operations

- Select the **Bore** option from the **Operation Type** drop-down in **PropertyManager** and set the other parameters as discussed earlier.
- Click on the **OK** button from the **PropertyManager**. The **Operation Parameters** dialog box will be displayed with options for bore holes; refer to Figure-52.

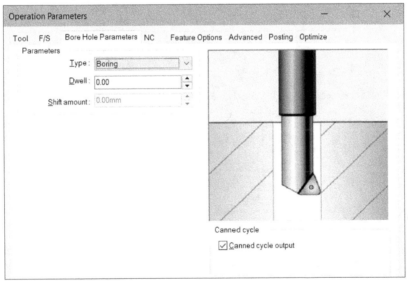

Figure-52. Operation Parameters dialog box with Bore hole options

- Select desired option from the **Type** drop-down to define type of boring operation. Select the **Boring** option from the drop-down to perform simple boring operation. Select the **Boring with Dwell** option from the drop-down to specify delay while boring for better heat management of cutting tool and easy removal of chips. Select the **Back Boring** option from the drop-down if you want to bore backside of hole with respect to machine headstock; refer to Figure-53. Select the **Fine Boring** option from the drop-down if you want to perform better surface finish holes. Select the **Counter Boring** option from the drop-down if you want to machine a counter bore hole.

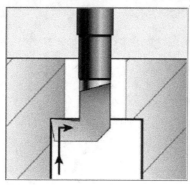

Figure-53. Back boring

- Set the other parameters as discussed earlier and click on the **OK** button from the dialog box to create operation.

Creating Reaming Operation

- Select the **Ream** option from the **Operation Type** drop-down if you want to finish drilled holes and set the other parameters in **PropertyManager** as discussed earlier.
- Click on the **OK** button from the **PropertyManager**. The **Operation Parameters** dialog box will be displayed with ream hole options; refer to Figure-54.

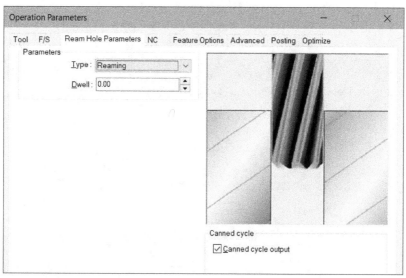

Figure-54. Operation Parameters dialog box with ream hole options

- Select the **Reaming** option from the **Type** drop-down to perform simple reaming operation in the holes. Select the **Ream w/dwell** option from the **Type** drop-down to perform reaming operation with specified delays.
- Set the other parameters as discussed earlier and click on the **OK** button to create the operation.

Creating Tap Operation

- Select the **Tap** option from the **Operation Type** drop-down if you want to perform threading using Tap cutting tool. Set the parameters in **PropertyManager** as discussed earlier and click on the **OK** button from the **PropertyManager** and **Properties** dialog box. The **Operation Parameters** dialog box will be displayed as shown in Figure-55.
- Select the **Tapping** option from the **Type** drop-down to create threading when tap is moving downward in material. Select the **Reverse Tapping** option from the **Type** drop-down to create threading when tap is moving upward in material.

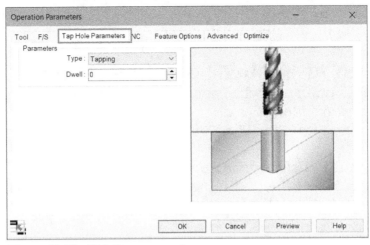

Figure-55. Tap Hole Parameters tab

- Specify desired value of pause in seconds between cutting passes when performing tap machining in the **Dwell** edit box.
- Set the other options as discussed earlier in the dialog box and click on the **OK** button to create operation.

The other options in **Operation Type** drop-down of **New Operation PropertyManager** for drilling have been discussed earlier.

3 AXIS MILL OPERATIONS

The 3 Axis mill operations are performed for machining parts while moving cutting tool simultaneously in X, Y, and Z axes. The procedure to create 3 axis mill operation is discussed next.

- Click on any tool from **3 Axis Mill Operations** cascading menu of the **SOLIDWORKS CAM CommandManager** in the **Ribbon**. The **New Operation PropertyManager** will be displayed with options for creating 3 axis milling operations; refer to Figure-56.

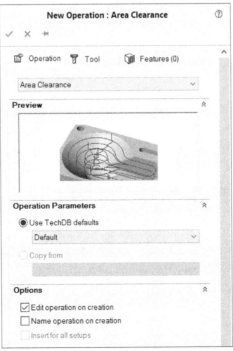

Figure-56. New Operation PropertyManager for 3 Axis Mill operations

- The options of Area Clearance operation have been discussed earlier. The other 3 Axis mill operations are discussed earlier.

Creating Z Level Operation

- Select the **Z Level** option from the **Operation Type** drop-down to machine contours while going downward along Z axis.
- Select desired option from the **Use TechDB defaults** drop-down to select strategy for operation.
- Select desired multi surface feature from the **Features** tab in **New Operation PropertyManager**. Select the cutting tool as discussed earlier and click on the **OK** button. The **Operation Parameters** dialog box will be displayed; refer to Figure-57.

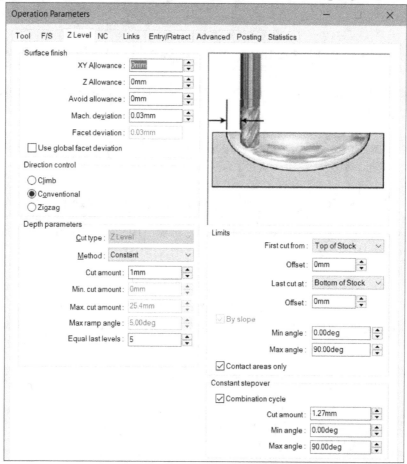

Figure-57. Operation Parameters dialog box with Z level options

- Specify desired parameters in the dialog box as discussed earlier and click on the **OK** button from the dialog box. The Z level operation will be created.

Creating Flat Area Operation

- Select the **Flat Area** option from the **Operation Type** drop-down in **New Operation PropertyManager**.
- Set the other parameters in the **PropertyManager** as discussed earlier and click on the **OK** button. The **Operation Parameters** dialog box will be displayed with flat area options; refer to Figure-58.

- Specify desired value in the **Number of cuts** edit box to define number of cutting passes for each Z level while machining.
- Specify desired value in **Axial offset** edit box to define distance between two consecutive axial cutting passes.

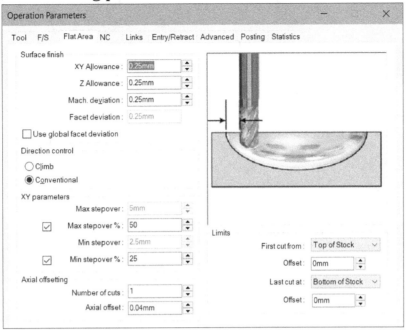

Figure-58. Flat Area options in Operation Parameters dialog box

- Specify the other parameters in dialog box as discussed earlier and click on the **OK** button to create operation.

DEFINING NEW AVOID AREA

The **New Avoid Area** tool is used to define areas in the model that should not get machined by current selected operation. The procedure to use this tool is given next.

- Select the operation for which you want to define avoid area from the **SOLIDWORKS CAM Operation Tree** and click on the **New Avoid Area** tool from **SOLIDWORKS CAM CommandManager** in the **Ribbon**; refer to Figure-59. The **Avoid Area PropertyManager** will be displayed; refer to Figure-60.

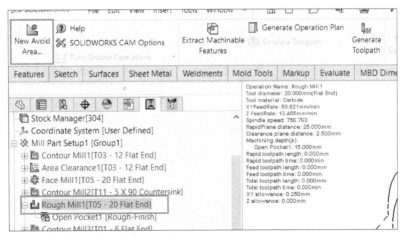

Figure-59. New Avoid Area tool

Figure-60. Avoid Area PropertyManager

- Select desired options from the **Face Selection** and **Edge Selection** drop-downs in the **Selection Mode** rollout of **PropertyManager**.
- Select desired faces and edges from the model to define a closed boundary avoid area where tool will not approach while performing machining operations.
- You can also select desired sketches from the **Allowed Sketches** rollout in the **PropertyManager**.
- Select desired button from the **Shape** rollout to define shape by which selected boundaries can be offset to increase the avoid area. After select the button from this rollout, specify desired value of offset distance in the **Offset** edit box.
- Specify desired value of heights/depths of avoid zone in **Direction 1** and **Direction 2** rollouts of **PropertyManager** to define upward and downward scope of avoid zone, respectively.
- Set other parameters in **PropertyManager** as discussed earlier and click on the **OK** button. The avoid area will be assigned to selected operation; refer to Figure-61.

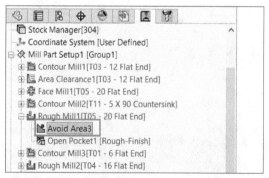

Figure-61. Avoid area assigned to operation

DEFINING NEW CONTAINMENT AREA

The **New Contain Area** tool is used to define containment area for selected operation. Defining containment area ensures that cutting tool will not move outside defined boundaries while cutting. The procedure to use this tool is given next.

- Select the operation for which you want to create containment area from the **SOLIDWORKS CAM Operation Tree** and click on the **New Contain Area** tool from the expanded **SOLIDWORKS CAM CommandManager** in the **Ribbon**; refer to Figure-62 or **Tools > SOLIDWORKS CAM > New > Contain Area** tool from menubar. The **Contain Area PropertyManager** will be displayed; refer to Figure-63.

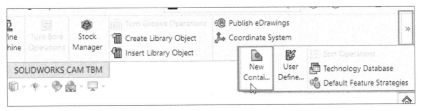

Figure-62. New Contain Area tool

Figure-63. Contain Area PropertyManager

- Select desired faces, edges, sketch boundaries as discussed earlier for **New Avoid Area** tool.
- Set desired parameters and click on the **OK** button from the **PropertyManager** to create the containment area.

PROBING

Probing systems on CNC machining centres and lathes can be used to identify and set-up parts, measure features in-cycle for adaptive machining, monitor workpiece surface condition and verify finished component dimensions. The procedure to create probing operation is given next.

- Click on the **Probe Operation** tool from the **SOLIDWORKS CAM CommandManager** in the **Ribbon**. The **Setup for Probe Operation PropertyManager** will be displayed; refer to Figure-64.

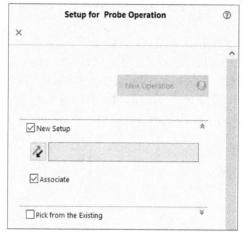

Figure-64. Setup for Probe Operation PropertyManager

- Select desired face to define direction of probing tool. You can select an existing setup from the **Pick from the Existing** rollout.
- Click on the **New Operation** button from the **PropertyManager**. The **Tool** tab of **New Operation PropertyManager** will be displayed; refer to Figure-65.

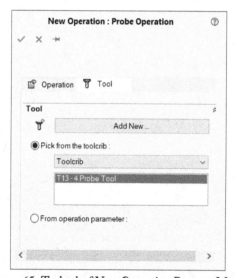

Figure-65. Tool tab of New Operation PropertyManager

- Select desired probe tool and click on the **OK** button from the **PropertyManager**. The **Operation Parameters** dialog box will be displayed with probe options; refer to Figure-66.

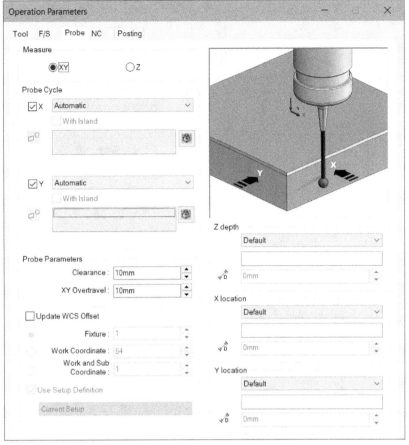

Figure-66. Probe tab of Operation Parameters dialog box

- Select desired radio button from the **Measure** area to define whether you want to measure XY coordinates or you want to measure Z coordinates.

- If you have selected **XY** radio button in **Measure** area then options for measuring X and Y coordinates will be displayed in the **Probe Cycle** area of the dialog box. If **Z** radio button is selected then options for measuring Z coordinate of single face.

- In our case, we have selected the **XY** radio button for describing options. Select desired option from the **X** drop-down to define what type of feature is to be measured. Select the **Single Face** option from the **X** drop-down if you want to move probe over single face for measuring X coordinates at different locations; refer to Figure-67. Select the **Web** option from the **X** drop-down if you want to measure the feature width and centre position using 2 points parallel to the X or Y axis; refer to Figure-68. Select the **Pocket** option from the drop-down if you want to measure width or diameter of a pocket. Select the **With Island** check box if there is an island in the pocket to be measured; refer to Figure-69. Select the **Boss** or **Bore** option from the drop-down if you want to measure the feature width and centre position using 4 points parallel with the X and Y axis, to determine the diameter size and centre position. The error of size and centre position can be stored or used to update the relevant work offset registers. Select the **Three Point Boss** or **Three Point Bore** option from the drop-down to measure boss or bore features using center and three angle positions; refer to Figure-70 and Figure-71.

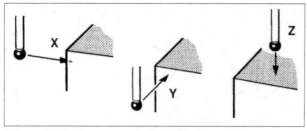

Figure-67. Probe movement for single face

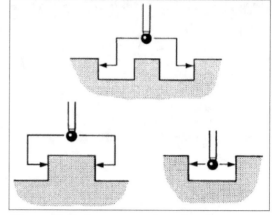

Figure-68. Probe movement for web cycle

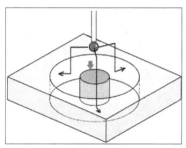

Figure-69. Probe movement for pocket cycle with islands

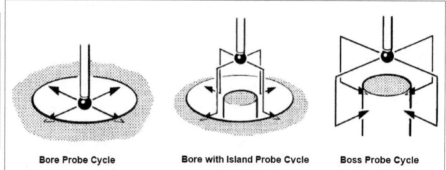

Bore Probe Cycle   **Bore with Island Probe Cycle**   **Boss Probe Cycle**

Figure-70. Probe movement for bore and boss cycles

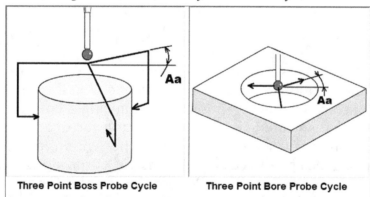

Three Point Boss Probe Cycle   **Three Point Bore Probe Cycle**

Figure-71. Probe movement for three point bore and boss cycles

- Select the face(s) to be measured by probe; refer to Figure-72.
- Specify desired value in the **Clearance** edit box to define distance from the part face before approaching to measure.
- Specify desired value in **XY Overtravel** edit box to define distance allowed for probe to travel past selected faces.
- Select the **Update WCS Offset** check box to update the work coordinate offset value based on points measured by probe.
- Clear the **Use Setup Definition** check box if you want to specify Fixture number, Work Coordinate, or Work and Sub Coordinate value by selecting respective radio button.
- Specify desired values in **Z depth**, **X location**, and **Y location** edit boxes to offset the measurement point from selected reference.

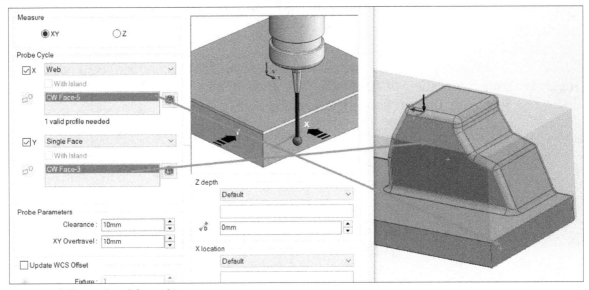

Figure-72. Faces selected for probing

Feed and Speed for Probing

- Click on the **F/S** tab in the dialog box to display feed and speed parameters for probing; refer to Figure-73.

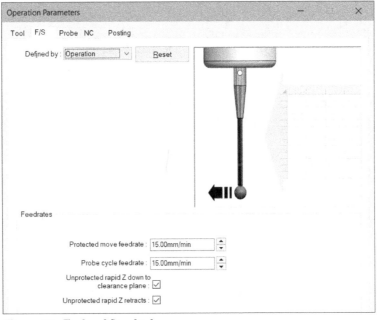

Figure-73. Feed and Speed tab

- Select desired option from the **Defined by** drop-down to define whether you want to use operation parameters specified in this dialog box or tool parameters for feed & speed of probe.
- If **Operation** option is selected in the drop-down then specify desired feed rates for protected moves and prove moves in respective edit boxes of the **Feedrates** area.
- Select the **Unprotected rapid Z down to clearance plane** check box if you want to move probe rapidly to clearance plane for unprotected Z down moves.
- Select the **Unprotected rapid Z retracts** check box if you want probe rapidly retract in Z direction.

- Specify the parameters in other tabs as discussed earlier and click on the **OK** button to create probing operation.

Note that you can create probe operation for a milling operation by right-clicking on the operation from the **SOLIDWORKS CAM Operation Tree PropertyManager** and selecting the **Probing Operation** option; refer to Figure-74.

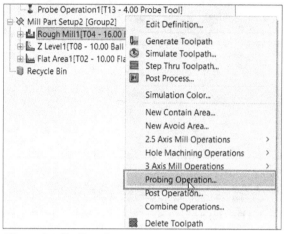

Figure-74. Probing Operation option

SELF ASSESSMENT

Q1. Select the **Lock spindle speed** check box if you want to keep RPM fixed even when diameter of tool has changed.

Q2. Which of the following options is used to define the movement speed at which cutting tool will enter the workpiece to begin machining?

a. Leadin feedrate edit box   b. Leadout feedrate edit box
c. XY feedrate edit box   d. Enable corner slowdown check box

Q3. Discuss the difference between Climb machining and Conventional machining.

Q4. What is the function of rapid plane in CAM programming?

Q5. The options in the **Posting** tab are used to define coolant type and NC code output type. (T/F)

Q6. The contour milling is used to finish walls of boss and pocket features. (T/F)

Q7. The thread mill operation is used to create internal and external threads. (T/F)

Q8. Select the **Spot Drilling** option if you want to drill a guide hole for other drilling operations. (T/F)

Chapter 5

Toolpath Generation and Processing

Topics Covered

The major topics covered in this chapter are:

- *Introduction.*
- *Generating Toolpath.*
- *Simulating Toolpaths.*
- *Step Thru Toolpath Simulation.*
- *Saving Controller File.*
- *Post Processing.*

INTRODUCTION

In previous chapter, you have learned to create various milling operation plans. The next step in SolidWorks CAM is to generate toolpaths based on milling operations. The tools and options related to milling toolpaths are discussed next.

GENERATING TOOLPATH

The **Generate Toolpath** tool is used to generate toolpaths for selected operations. The procedure to use this tool is given next.

- Select desired mill part setup from the **SOLIDWORKS CAM Operation Tree** if you want to use all the operations in selected part setup for generating toolpath or select an operation for generating single toolpath. Click on the **Generate Toolpath** button from the **SOLIDWORKS CAM CommandManager** in the **Ribbon**. The **SOLIDWORKS CAM Process Manager** will be displayed; refer to Figure-1.

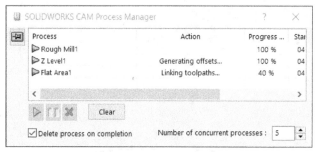

Figure-1. SOLIDWORKS CAM Process Manager

- Once the process of generating toolpath is complete, you can check the toolpath by selecting it from the **SOLIDWORKS CAM Operation Tree**; refer to Figure-2.

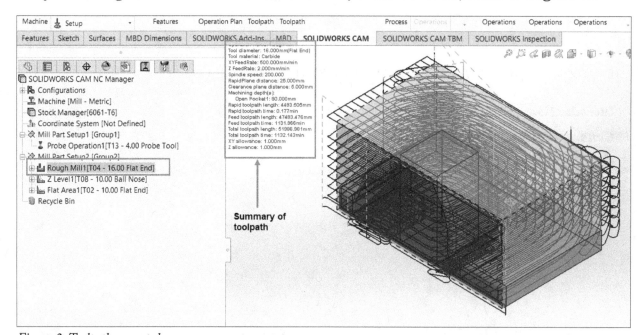

Figure-2. Toolpath generated

You can check the details of toolpath like tool diameter, rapid toolpath length, feed toolpath length, and so on in the drawing area of interface.

SIMULATING TOOLPATHS

The **Simulate Toolpath** tool is used to perform simulation of cutting tool movements while performing the operation. The procedure to use this tool is given next.

- After selecting desired mill part setup or operation from the **SOLIDWORKS CAM Operation Tree** and click on the **Simulate Toolpath** tool from the **SOLIDWORKS CAM CommandManager** in the **Ribbon**. The **Simulate Toolpath PropertyManager** will be displayed with preview of NC model; refer to Figure-3.

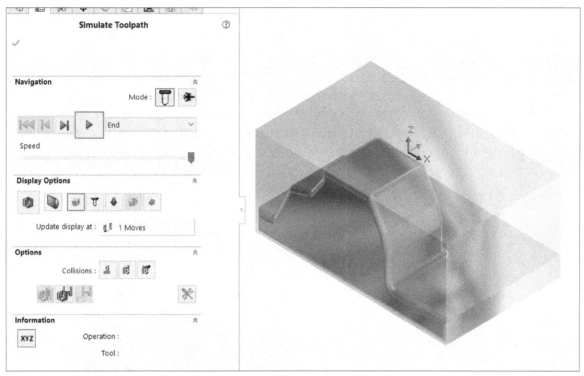

Figure-3. Simulate Toolpath PropertyManager

Navigation Rollout Options

- Select desired button from the **Mode** area of **Navigation** rollout in **PropertyManager** to define how toolpath will simulate. Select the **Tool Mode** button to view the cutting tool as it goes over toolpath through material while machining. The simulation in this mode is relatively slow and takes more computing resources. So, this mode is good for small workpieces with less number of toolpaths. Select the **Turbo Mode** button to handle large number of toolpaths with speed 10-20 times faster than tool mode. Note that some features will not be available in Turbo mode as compared to Tool mode like cutting tool will not display in simulation, you can simulate one setup at a time, if you are machining assembly then only one part machining will be displayed in simulation, fixtures will not be displayed in simulation, and so on.
- Using the buttons in **Navigation** rollout, you can run, single step forward, single step backward, and go to start position of simulation.
- Select desired option from the drop-down in the **Navigation** rollout to define upto what level you want to perform simulation. Select the **Next Z Level** option from the drop-down if you want to perform simulation upto the point where cutting tool will move to next Z level depth. Select the **Next Toolpath** option from the drop-down if you want to stop simulation when tool reaches next toolpath. Select the **Next Operation** option from the drop-down if you want to stop simulation when toolpath for current operation is complete. Select the **Next Tool** option from the

drop-down if you want to stop the simulation when cutting tool change occurs. Select the **End** option from the drop-down if you want to run simulation upto the end of current mill setup.

*   Using the slider in **Navigation** rollout, you can set the speed of simulation.

Display Options Rollout

The options in **Display Options** rollout are used to set parameters for displaying various objects in the simulation.

*   Select the **Show Difference** toggle button to show part, workpiece stock, and machined stock in difference colors.
*   Click on the **Section view** button to check section of part and workpiece while machining. The **Section View** dialog box will be displayed with preview of model; refer to Figure-4. Select desired button from **Section Plane** area to use XY, YZ, or XZ plane for section. Click on the **Reverse direction** button to flip the direction of section. Specify desired positive or negative value in the **Offset** edit box to move section plane at specified distance from the plane. Click on the **Close** button or click in the empty area of drawing to exit the dialog box.

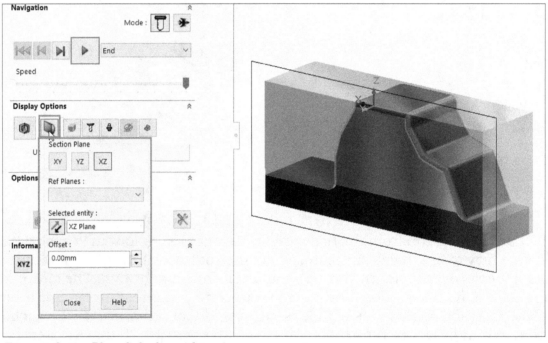

Figure-4. Section Plane dialog box with preview

*   Click on the **Stock Display** button to select desired stock display options; refer to Figure-5. Select **No Display** option if you do not want to display stock. Select the **Wireframe Display** option if you want to display stock in wireframe style. Select the **Translucent Display** option if you want to transparent stock. Select the **Shaded Display** option if you want to display stock as shaded. Select the **Shaded With Edges** option if you want to display stock shaded with edges as dark lines.

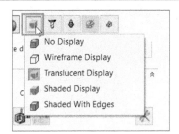

Figure-5. Stock display options

- Similarly, click on the **Tool Display** button and **Tool Holder Display** button to change display style of cutting tool and tool holder respectively.
- If you are working on assembly for machining then click on the **Fixture** button to change display style for fixtures.
- Click on the **Target Part Display** button to change display style of part model used for machining reference.
- Set desired interval at which display of model will be updated during simulation in the **Update display at** drop-down of **Display Options** rollout.

Options Rollout

- Click on the **Tool Cut Collision** button from the **Collisions** area of **Options** rollout to specify whether to stop simulation when cutting tool collides with stock or cut through even on collision; refer to Figure-6. Select the **Ignore Collision** option if you want cutting tool to pass through stock during simulation. Select the **Cut Collision** option if you want the cutting tool to cut through stock in case of collision. Select the **Pause on Collision** option if you want to stop simulation in case of tool collision with stock/fixture/part.

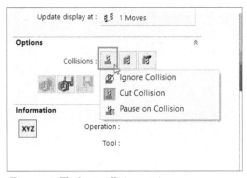

Figure-6. Tool cut collision options

- Similarly, you can set options for tool's non cutting edges and tool holder during collision by selecting the **Tool Shoulder/Shank** and **Tool Holder** buttons, respectively.
- Once the simulation is complete, select the chips left after machining and press **CTRL+D** to remove them.
- Click on the **Save WIP as STL** button from the **Options** rollout if you want to save current model after performing machine simulation as workpiece in progress for next operation in STL format. The **Save As** dialog box will be displayed; refer to Figure-7. Specify desired name, unit, output coordinate system, and so on in the dialog box. Click on the **Save** button to save the file.

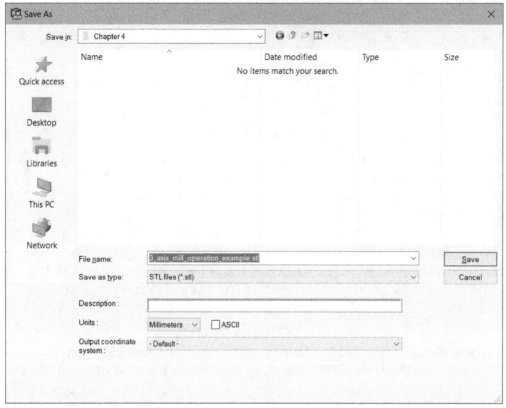

Figure-7. Save As dialog box

- Click on the **Save .jpg image for Setup sheets** button to save current model state as image file. The **Save WIP Image As** dialog box will be displayed; refer to Figure-8.

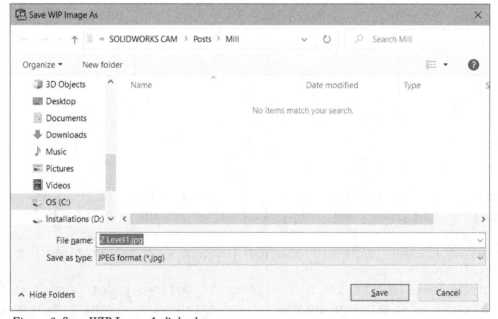

Figure-8. Save WIP Image As dialog box

- Specify desired name and format for the image file, and click on the **Save** button to save the image files.

Specifying Simulation Options

Click on the **Options** button from the **Options** rollout of **Simulate Toolpath PropertyManager**. The **Options** dialog box will be displayed; refer to Figure-9. Various options in this dialog box are discussed next.

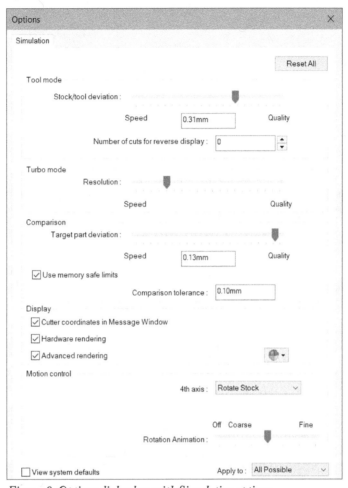

Figure-9. Options dialog box with Simulation options

- Move the slider of **Stock/tool deviation** to define the tolerance within which tool and stock representations can vary from original.
- Specify desired value in the **Number of cuts for reverse display** edit box to define for how many cutting steps, the material can be added while you are performing reverse step simulation.
- Move the **Resolution** slider in **Turbo mode** area to desired side for defining how good the model should be displayed when running simulation in turbo mode.
- Move the **Target part deviation** slider in **Comparison** area to define tolerance for displaying part in simulation.
- Select the **Use memory safe limits** check box to make sure the memory usage by simulation are under safe limits.
- Select the **Cutter coordinates in Message Window** check box to display coordinates of cutting tool during simulation in message window.
- Select the **Hardware rendering** check box to use graphics card support for simulation rendering.
- Select the **Advanced rendering** check box to define material type for various components in simulation. Click on the **Advanced rendering** button next to check box and select material for different components; refer to Figure-10.

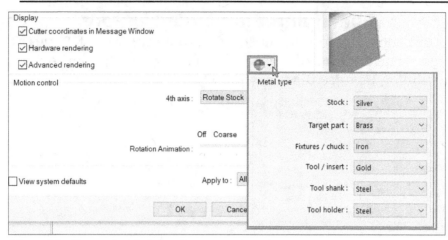

Figure-10. Advanced rendering options

- Select desired option from the **4th axis** drop-down in the **Motion control** area of the dialog box. Select the **Rotate Stock** option if the stock can rotate in machine. Select the **Rotate Tool** option if the cutting tool can rotate/tilt in machine.
- Set desired simulation quality for rotation in **Rotation Animation** slider of **Motion control** area.
- Select desired option from the **Apply to** drop-down to define scope where you want to apply the settings.
- Click on the **OK** button from the dialog box to apply settings.

STEP THRU TOOLPATH SIMULATION

The **Step Thru Toolpath** tool is used to display movement of cutting tool over toolpath during simulation. You can also check the spindle rpm, feedrate, and other parameters in real-time during simulation. The procedure to use this tool is given next.

- Click on the **Step Thru Toolpath** tool from the **SOLIDWORKS CAM CommandManager** in the **Ribbon**. The **Step Through Toolpath PropertyManager** will be displayed; refer to Figure-11.
- Set desired speed for tool movement in the **Speed** slider.
- Set the display styles for various components in **Display Options** rollout as discussed earlier.

Figure-11. Step Through Toolpath PropertyManager

- Click on the **Play/Pause** button from the **Navigation** rollout to start or stop toolpath simulation. You can pause the simulation anytime by clicking on the button again. The simulation of toolpath will be displayed as shown in Figure-12.

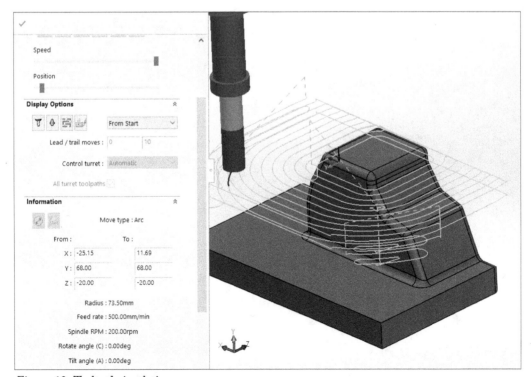

Figure-12. Toolpath simulation

- After checking the simulation, click on the **OK** button from **PropertyManager** to exit.

SAVING CONTROLLER FILE

The **Save CL File** tool is used to save toolpath in the form of CL file which can be analyzed by text editors or post processors. Using the post processor, you can convert the CL file into CNC controller specific file. Note that this option may not be available in educational version. The procedure to save CL file is given next.

- Click on the **Save CL File** tool from the **SOLIDWORKS CAM CommandManager** in the **Ribbon**. The **Save As** dialog box will be displayed; refer to Figure-13.

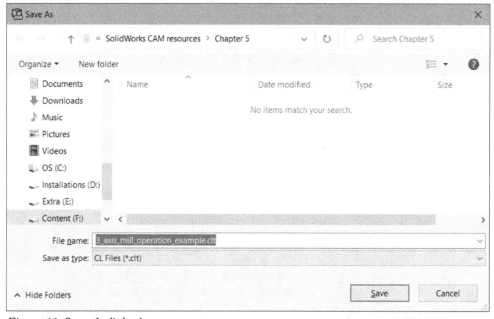

Figure-13. Save As dialog box

- Specify desired name of CL file in the **File name** edit box and click on the **Save** button. The file will be saved in specified location. Double-click to open the file; refer to Figure-14.

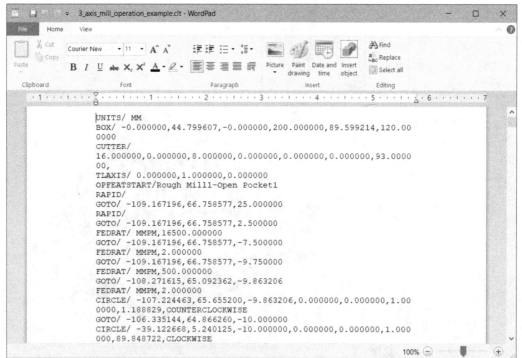

Figure-14. CL file in WordPad

POST PROCESSING

The **Post Process** tool is used to convert CL file into machine controller specific codes for your CNC machine. Note that this tool will be active only if you have defined post processor for your machine in **Machine** dialog box as discussed earlier. The procedure to use this tool is given next.

- Once the toolpaths have been generated and you have checked the CL file, click on the **Post Process** tool from the **SOLIDWORKS CAM CommandManager** in the **Ribbon**. The **Post Output File** dialog box will be displayed; refer to Figure-15.

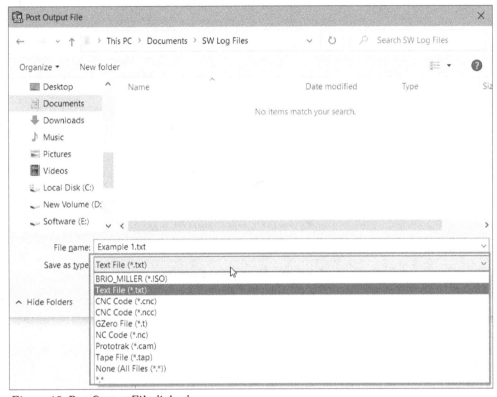

Figure-15. Post Output File dialog box

- Specify desired name of file in the **File name** edit box and select desired format for file.
- Move to desired location in the dialog box and click on the **Save** button to save the file. The **Post Process PropertyManager** will be displayed; refer to Figure-16.
- Select the **Centerline** check box from the **Options** rollout to display the toolpath centerlines while generating codes.
- Select the **Open G-Code file in** check box to open the generated G codes in SOLIDWORKS CAM NC Editor application.
- After setting desired parameters, click on the **Play** button to run the code generation. Once the code generation is complete, the codes will be displayed in the **NC Code** rollout of **PropertyManager**.
- Click on the **OK** button from the **PropertyManager**. The **SOLIDWORKS CAM NC Editor** application window will be displayed; refer to Figure-17.

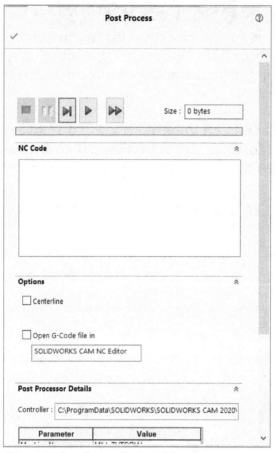

Figure-16. Post Process PropertyManager

Figure-17. SOLIDWORKS CAM NC Editor

- Modify the codes as desired in the editor and save the file. You can directly use this file in your machine via USB or other file transfer drive.

SORTING OPERATIONS

The **Sort Operations** tool is used to change the order of operations in the NC program based on specified parameters. The procedure to use this tool is given next.

- Select the mill part setup whose operations are to be sorted and click on the **Sort Operations** tool from the **Ribbon**. The **Sort Operations** dialog box will be displayed; refer to Figure-18.

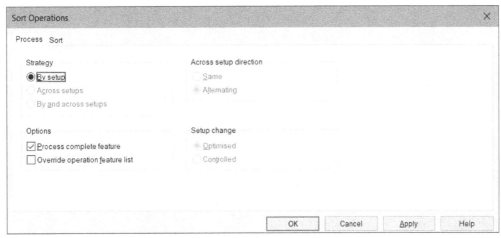

Figure-18. Sort Operations dialog box

- Set desired options from the **Process** tab to define what will be included in the sorting process.
- Select the **Sort** tab to define how sorting of operations will be performed. The options will be displayed as shown in Figure-19.

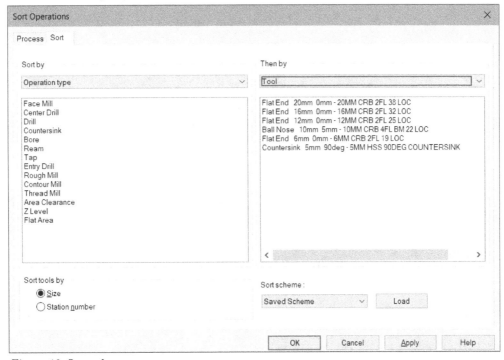

Figure-19. Sort tab

- You can drag the options in the lists to define custom order of operations.
- After setting desired parameters, click on the **OK** button. The operations will be sorted automatically.

FOR STUDENT NOTES

Chapter 6

Tolerance Based Machining and Miscellaneous Tools

Topics Covered

The major topics covered in this chapter are:

- *Introduction.*
- *Tolerance Based Machining*
- *Technology Database*
- *User Defined Tool/Holder*
- *Creating and Inserting Library Objects*
- *Publishing eDrawings*
- *Help*

INTRODUCTION

In previous chapter, you have learned to create to generate toolpaths based on milling operations. You have also learned to generate NC output files for machines. In this chapter, you will learn to perform tolerance based machining. You will also learn about various miscellaneous tools like creating user defined tools/holders, creating library objects, and so on. These tools are discussed next.

TOLERANCE BASED MACHINING

The tools to perform tolerance based machining are available in **SOLIDWORKS CAM TBM CommandManager** in the **Ribbon** and **SOLIDWORKS CAM TBM** toolbar. To display **SOLIDWORKS CAM TBM** toolbar, right-click on any tab in the **Ribbon** and select the **SOLIDWORKS CAM TBM** option from **Toolbar** cascading menu in the shortcut menu; refer to Figure-1. The **SOLIDWORKS CAM TBM** toolbar will be displayed; refer to Figure-2. The **Tolerance Based Machining - Run** tool is used to generate machining features based on tolerance condition of features in the model. Before using this tool, make sure your model has tolerances applied to it; refer to Figure-3. After that we apply settings related to tolerance based machining using the **Tolerance Based Machining (Mill) - Settings** tool. The procedure to define settings is given next.

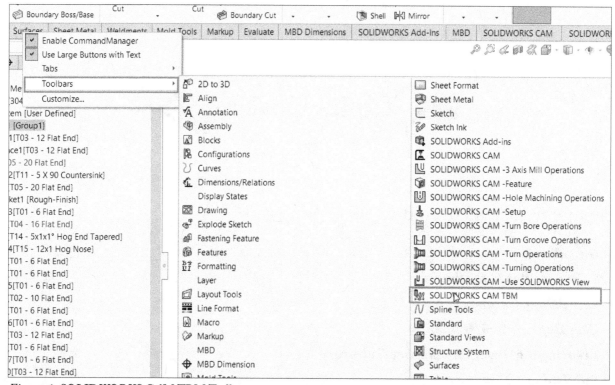

Figure-1. SOLIDWORKS CAM TBM Toolbar

Figure-2. SOLIDWORKS CAM TBM Toolbar

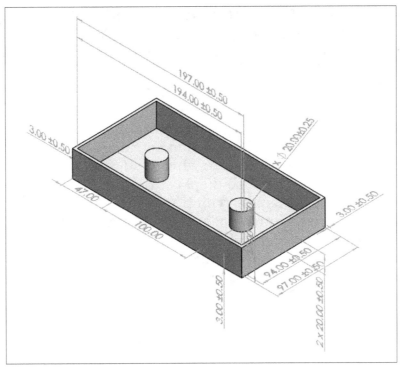

Figure-3. Model with tolerances applied

Defining Tolerance Based Machining Settings

- Click on the **Tolerance Based Machining (Mill) - Settings** tool from the **SOLIDWORKS CAM TBM CommandManager** in the **Ribbon** or **SOLIDWORKS CAM TBM** toolbar. The **SOLIDWORKS CAM Tolerance Based Machining(Mill) - Settings** dialog box will be displayed; refer to Figure-4.

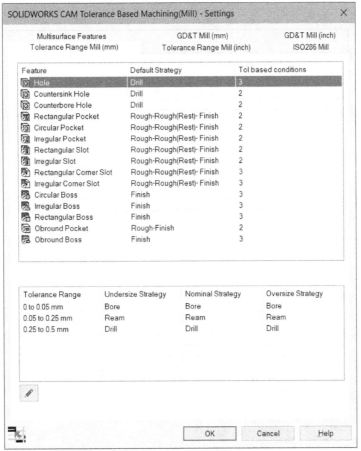

Figure-4. SOLIDWORKS CAM Tolerance Based Machining(Mill) - Settings dialog box

Modifying Tolerance Range and Strategy

- The options in the **Tolerance Range Mill(mm)** tab of dialog box are used to set tolerance range and related machining strategies. For example, the hole feature in the table has default strategy as drilling but there are three conditions specified for tolerance. When tolerance for hole diameter is between 0 to 0.05 mm then bore strategy will be applied, when hole diameter tolerance is between 0.05 to 0.25 mm then reaming strategy will be used for machining, and when tolerance range is between 0.25 to 0.5 mm then default drilling strategy will be used for machining.

- Select desired feature from the table for which you want to set tolerance range and strategy, the related options will be displayed in the bottom table of dialog box.

- Select desired tolerance range for which you want to define strategy and then set desired options in respective **Undersize Strategy**, **Nominal Strategy**, and **Oversize Strategy** fields; refer to Figure-5.

- If you want to modify the tolerance range then click on the **Edit Tolerance Range** button at the bottom in the dialog box. The **Range** dialog box will be displayed; refer to Figure-6.

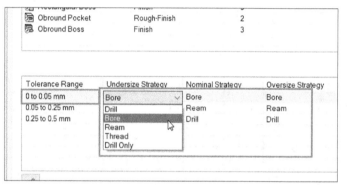

Figure-5. Setting strategies for tolerance ranges

Figure-6. Range dialog box

- Specify desired value in the edit box and click on the **Add** button. A new range will be added. If you want to delete any tolerance range then select it from the table in the **Range** dialog box and press **DELETE** button from keyboard.

- Click on the **OK** button from the dialog box. The tolerance range will be modified accordingly; refer to Figure-7. You can set the strategies as discussed earlier.

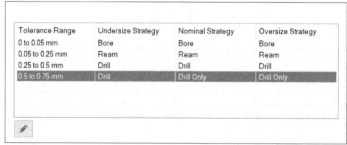

Figure-7. Modified tolerance range

Similarly, you can set the tolerance ranges in Inches using the options in the **Tolerance Range Mill (inch)** tab of the dialog box; refer to Figure-8.

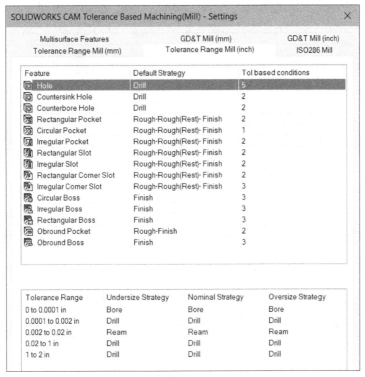

Figure-8. Tolerance Range Mill Inch options

ISO286 Mill Options

The options in **ISO286 Mill** tab are used to define tolerance conditions for holes and circular boss (shaft) features; refer to Figure-9. You can modify the options in this tab as discussed earlier.

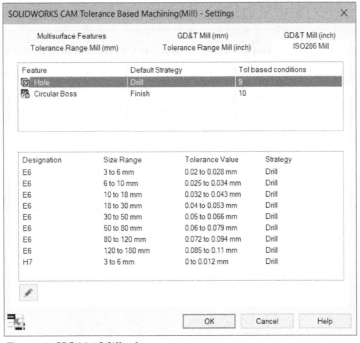

Figure-9. ISO286 Mill tab

GD&T Mill (mm) Options

The options in the **GD&T Mill (mm)** tab are used to define tolerance range and machining strategy for various geometric dimensioning and tolerance symbols;

refer to Figure-10. You can define strategy for new symbols by clicking on the **Edit Tolerance Range** button at the bottom in the dialog box as discussed earlier.

Figure-10. GD&T Mill tab options

You can specify the parameters in **GD&T Mill (inch)** tab in the same way.

Multisurface Features Options

The options in the **Multisurface Features** tab are used to define surface finish range and 3 axis mill machining strategy for the multisurface parts; refer to Figure-11. Set desired strategies for various surface finish ranges in the table as discussed earlier.

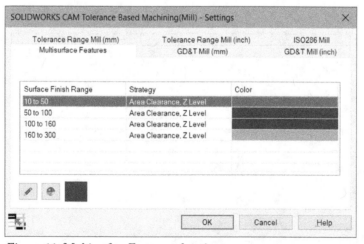

Figure-11. Multisurface Features tab options

- After setting desired parameters in the dialog box, click on the **OK** button.

Running Tolerance Based Machining

The **Tolerance Based Machining - Run** tool is used to recognize features based on their tolerance values and apply respective machining strategies. The procedure to use this tool is given next.

- Click on the **Tolerance Based Machining - Run** tool from the **SOLIDWORKS CAM TBM CommandManager** in the **Ribbon**. The **SOLIDWORKS CAM Tolerance Based Machining(Mill) - Run** dialog box will be displayed; refer to Figure-12.

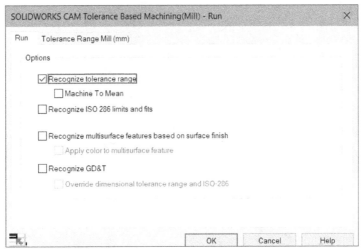

Figure-12. SOLIDWORKS CAM Tolerance Based Machining(Mill) – Run dialog box

- Select the **Recognize tolerance range** check box if you want to use dimension tolerances applied to model for recognizing machinable features and apply machining strategies.
- Select the **Machine To Mean** check box to automatically calculate mean value of tolerances for various dimensions of the model. This option is mainly applicable for contour mill operations.
- Select the **Recognize ISO 286 limits and fits** check box if you want to include features which have limits and fits applied to them based on ISO 286. The hole base fits are applied as H7/h6, H6/k5, and so on. The shaft base fits are applied as h6/G7, h6/H6, and so on.
- Select the **Recognize multisurface features based on surface finish** check box to include multisurface features based on parameters specified in the **SOLIDWORKS CAM Tolerance Based Machining(Mill) - Settings** dialog box. Select the Apply color to multisurface feature check box if you want to apply colors to model surfaces based on tolerance settings.
- Select the **Recognize GD&T** check box if you want to include features for machining based on geometric dimensioning and tolerances applied to the model. If you want to override the tolerances applied using ISO 286 by GD&T applied to the model.
- Note that based on selected check boxes, the respective tabs will be displayed in the dialog box. The options in these tabs are same as discussed earlier.
- After setting desired parameters, click on the **OK** button. The features will be extracted with strategies applied.

You can generate operation plans and toolpaths for these features as discussed earlier.

CREATING USER DEFINED TOOL/HOLDER

The **User Defined Tool/Holder** tool is used to create cutting tool/holder using the 3D model of tool/holder created in the SolidWorks. Note that the model must have a coordinate system applied to define 0,0 point of tool; refer to Figure-13. The procedure to use this tool is given next.

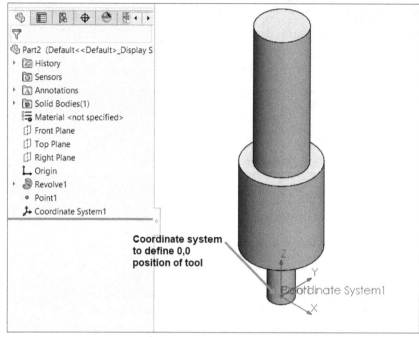

Figure-13. 3D model for creating tool

- After opening the model in SolidWorks, click on the **User Defined Tool/Holder** tool from the expanded **SOLIDWORKS CAM CommandManager** in the **Ribbon**; refer to Figure-14. The **Create Tool/Holder PropertyManager** will be displayed; refer to Figure-15.

Figure-14. User Defined Tool/Holder tool

- Select desired option from the **File Type** drop-down to define whether you want to create tool holder, mill tool, or turn insert.
- Click on the **Browse** button from the **Save To** rollout and specify the location where you want to save the user defined tool/holder file.
- After setting desired parameters, click on the **OK** button from **PropertyManager**. The mill tool/holder will be created.

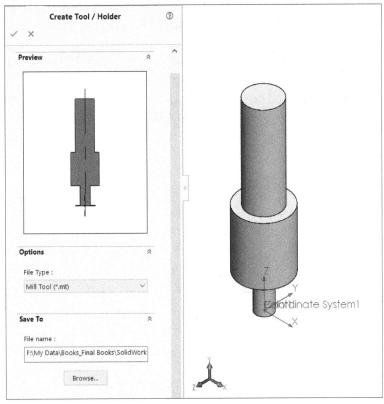

Figure-15. Create Tool Holder PropertyManager

CREATING LIBRARY OBJECTS

The **Create Library Object** tool is used to save SolidWorks CAM features and operations in library so that they can be used again later. Note that this tool will only be available if there are mill features and operations created in the model. The procedure to use this tool is given next.

- Select the feature which you want to add in library and click on the **Create Library Object** tool from the expanded **SOLIDWORKS CAM CommandManager** in the **Ribbon**. The **Add to Library** dialog box will be displayed; refer to Figure-16.

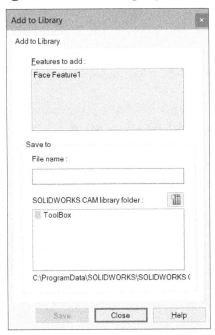

Figure-16. Add to Library dialog box

- You can select more features to be added to the library from the **SOLIDWORKS CAM NC Tree**. The selected features will be displayed in the **Features to add** selection box of dialog box.
- Specify desired name of file in the **File name** edit box of **Save to** area of dialog box.
- Select desired library folder in which you want to save the features from the **SOLIDWORKS CAM library folder** area of the dialog box. If you want to add a new library folder then click on the **Add a SOLIDWORKS CAM Library folder** button from the **Save to** area. The **Browse for Folder** dialog box will be displayed; refer to Figure-17.

Figure-17. Browse for Folder dialog box

- Select desired location and click on the **OK** button. The new location will be added to the list.
- After setting desired parameters, click on the **Save** button from the **Add to Library** dialog box.

INSERTING LIBRARY OBJECTS

The **Insert Library Object** tool is used to insert the objects earlier saved as CAM library objects by using the **Create Library Object** tool. The procedure to use this tool is given next.

- Click on the **Insert Library Object** tool from the **SOLIDWORKS CAM CommandManager** in the **Ribbon**. The **SOLIDWORKS CAM Library File** dialog box will be displayed; refer to Figure-18.
- Select desired file to be inserted in CAM model and click on the **Open** button. The **Add to Library** dialog box will be displayed with preview of machining features; refer to Figure-19.
- Select the feature for which you want to change the parameters from the **Machinable features** area and set the related parameters in **Geometry references** area of the dialog box.
- After setting desired parameters, click on the **Insert** button from the dialog box. The **Add to Library** dialog box will be displayed again after inserting features. Click on the **Cancel** button to exit the tool.

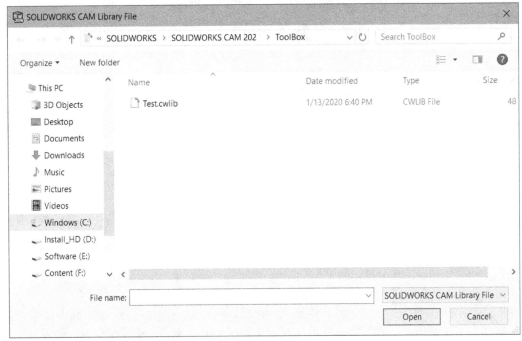

Figure-18. SOLIDWORKS CAM Library File dialog box

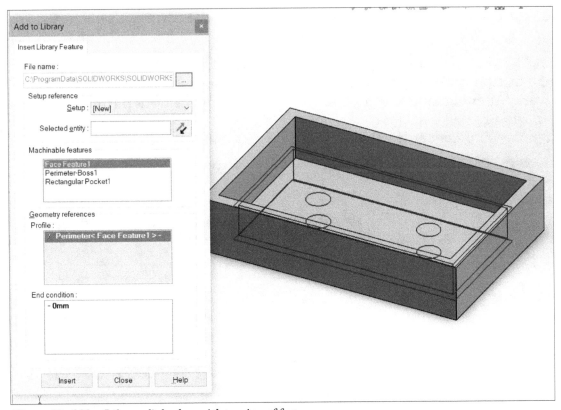

Figure-19. Add to Library dialog box with preview of features

PUBLISHING EDRAWING

The **Publish eDrawings** tool is used to create eDrawing of the model displayed in application. The procedure to use this tool is given next.

- Click on the **Publish eDrawings** tool from the **SOLIDWORKS CAM CommandManager** in the **Ribbon**. The model will be displayed in eDrawing application; refer to Figure-20.

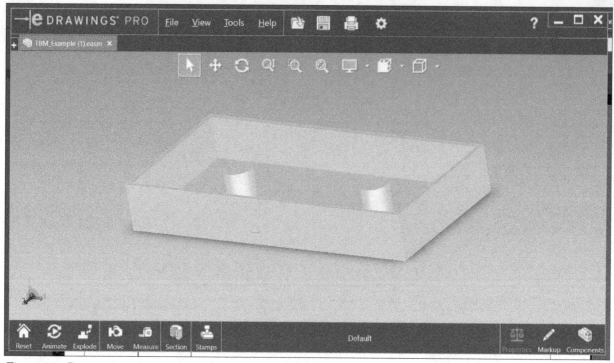

Figure-20. eDrawings application window

- You can use this file to share with others using eDrawings application.

SOLIDWORKS CAM HELP

The **Help** button in the top right corner of the **Ribbon** is used to access help database of software. After clicking this tool, select the **Help Topics** option from the **SOLIDWORKS CAM** help window will be displayed; refer to Figure-21. You can browse through various help topics using **Contents**, **Index**, and **Search** tabs of the dialog box.

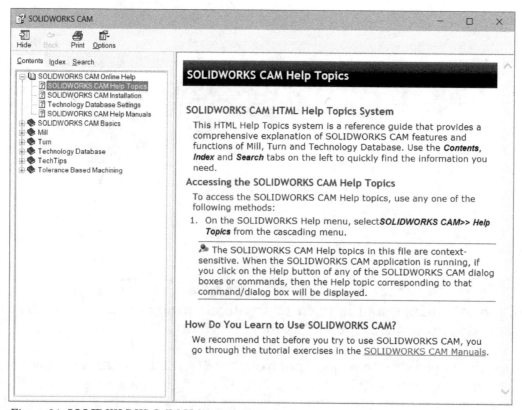

Figure-21. SOLIDWORKS CAM Help window

Chapter 7

Turning Operations

Topics Covered

The major topics covered in this chapter are:

- *Introduction.*
- *Defining Turning Machine.*
- *Defining Stock for Part.*
- *Turn Feature Options.*
- *Extracting Machinable Features.*
- *Creating Turn Features Manually.*
- *Generating Operation Plans Automatically and Manually.*

INTRODUCTION

In previous chapters, you have learned the procedures to perform milling operations. In this chapter, you will setup a turning machine for CAM and then generate various turning operations for machining.

DEFINING TURNING MACHINE

The **Define Machine** tool is used to set machine for performing CAM operations. This tool has been discussed earlier. Here, we will set the turning machine for CAM.

- Click on the **Define Machine** tool from the **SOLIDWORKS CAM CommandManager** in the **Ribbon**. The **Machine** dialog box will be displayed.
- Select desired turning machine from the **Turn Machines** node of **Available machines** area in the **Machine** tab of dialog box. After selecting the machine, click on the **Select** button.
- Set desired option in the **Machine duty** drop-down to define level of machining.
- Click on the **Tool Crib** tab to define turning tools and inserts; refer to Figure-1. These tools and inserts have been discussed earlier in Chapter 2.

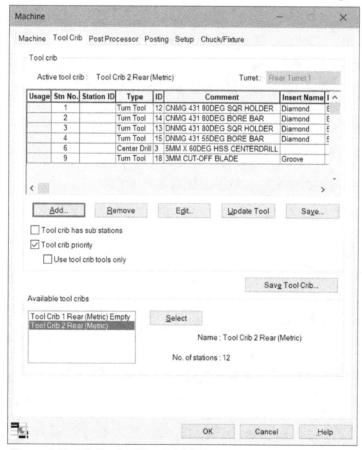

Figure-1. Tool Crib tab of Turning Machine

- Similarly, set the parameters in the **Post Processor** and **Posting** tab as discussed earlier.

Setup Options

- Click on the **Setup** tab to define axes and cutting planes for machining; refer to Figure-2.

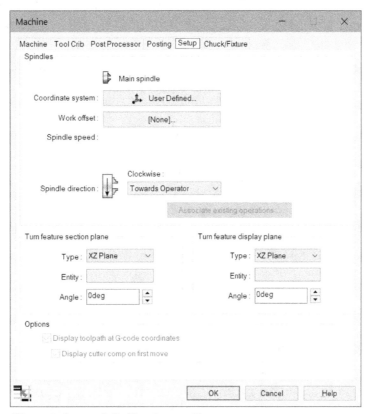

Figure-2. Setup tab for Turning machine

- Click on the **User Defined** button from the **Main spindle** area to change position of coordinate system for spindle location. Generally, the coordinate system is placed on the front face which is parallel to spindle jaws. On clicking this button, the **Main spindle coordinate systems PropertyManager** will be displayed; refer to Figure-3.
- Select the **Entity** radio button and select desired location to place the coordinate system; refer to Figure-4. You can change the direction of X, Y, or Z axis by selected references in respective selection boxes.
- After setting desired parameters, click on the **OK** button.

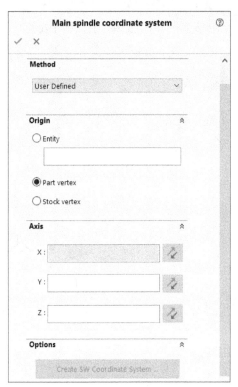

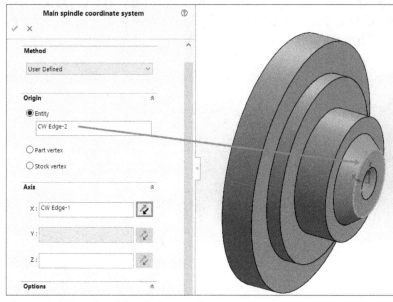

Figure-4. Edge selected for placing coordinate system

Figure-3. Main spindle coordinate system
PropertyManager

Defining Work offset for Turning

The work offset value is used to alter some of the cutting operations with specified minimal values. When we perform machining repeatedly with a cutting tool, there is always small wear in tool tip due to which cutting tool does not make cut enough to achieve desired dimensions. In such cases, we compensate this wear by specifying additional depth in cuts for concerned cutting tool.

- Click on the **[None]** button from the **Work offset** area of the dialog box. The **Spindle work coordinate** dialog box will be displayed; refer to Figure-5.

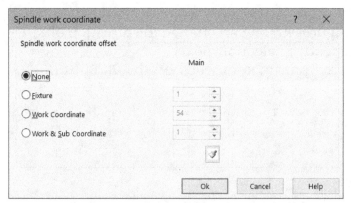

Figure-5. Spindle work coordinate dialog box

- Select the **None** radio button if you do not want to specify work offsets. Select the **Fixture** radio button if your machine supports programmable offsets in the format of E1, E2, and so on. This type of format is generally found in machines withe Delta controllers. Select the **Work Coordinate** radio button from the dialog box if your machine supports programmable offsets in the format of G54. G55, and so on. This format is generally supported in machines with Fanuc controllers.

Select the **Work & Sub Coordinate** radio button if your machine supports work coordinates with their sub categories like G54.1, G54 J1, and so on.

- After selecting the radio button, set desired value in respective spinner to assign offset number and click on the **OK** button from the dialog box.
- Set desired option in the **Spindle direction** drop-down to define rotation direction for spindle.
- The options in the **Turn features section plane** area are used to define the plane by which the part will be sectioned to automatically recognize the features. Select desired option from the **Type** drop-down to define plane. You can also select a reference face/plane to define the section plane after selecting the **Plane/Face** option from the **Type** drop-down. Click in the **Angle** edit box to define the angle by which the section plane will be rotated. Note that this section plane will be useful only when the **Revolved Section** option is selected as feature extraction method in **Options** dialog box; refer to Figure-6.

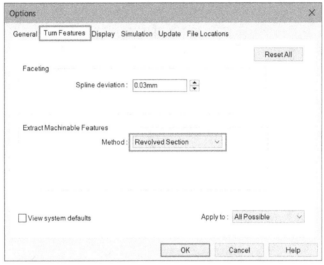

Figure-6. Revolved Section feature extraction method

- Set desired parameters in the **Turn features display plane** area to define at which plane the turn features will be displayed during simulation.
- Select the **Limit main spindle speed** check box to limit rotation of spindle at maximum specified value for all the operations.

Chuck/Fixture Options

The options in the **Chuck/Fixture** tab are used to define the shape and size of chuck/fixture used to hold the work piece; refer to Figure-7.

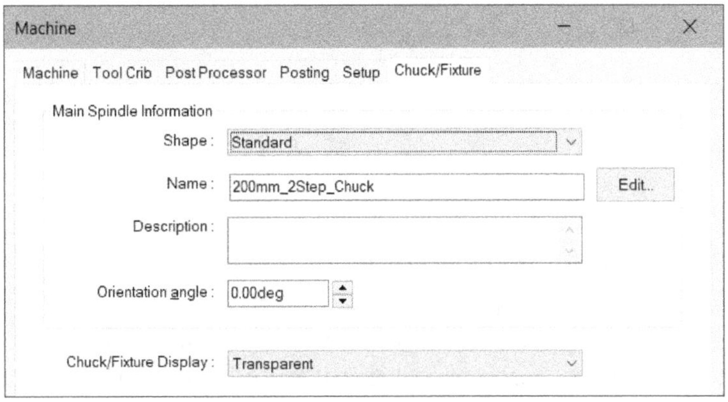

Figure-7. Chuck Fixture tab

- Select the **Standard** option from the **Shape** drop-down to use a cylindrical standard chuck for hold work piece; refer to Figure-8. To change the size of chuck or jaws, click on the **Edit** button next to **Name** field in the dialog box. The **Chuck Parameter : [Main Spindle] PropertyManager** will be displayed; refer to Figure-9.

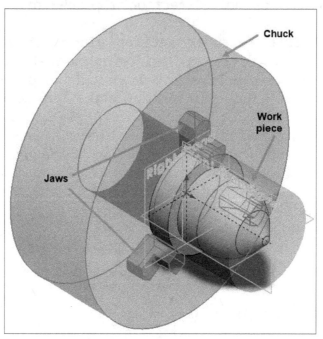

Figure-8. Preview of chuck

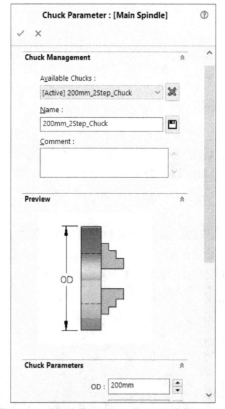

Figure-9. Chuck Parameter PropertyManager

- Select desired option from the **Available Chucks** drop-down in the **Chuck Management** rollout to use a predefined chuck for setup.

- If you want to change the size of chuck manually then specify desired values in **Chuck Parameters** rollout to define outer diameter, inner diameter, and thickness of chuck; refer to Figure-10.

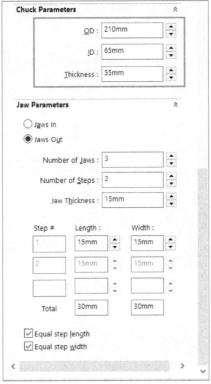

Figure-10. Chuck Parameters rollout

- If you have hollow workpiece then you can select the **Jaws In** radio button so that jaws hold the workpiece using inner diameter of stock. Select the **Jaws Out** radio button to hold the workpiece using outer diameter of stock. Specify the parameters related to jaw shape and size in the **Jaw Parameters** rollout like number of jaws, number of steps, and jaw thickness.
- If you want to change display style of chuck/fixture then select desired option from the **Chuck/Fixture Display** drop-down.
- If you want to use a SolidWorks part/assembly file or STL file then select respective option from the **Shape** drop-down and click on the **Browse** button next to drop-down for selecting the file.
- Click on the **OK** button from the dialog box to apply machine settings.

Note that the chuck created in model would represent the real machine chuck so that cutting tool can avoid collision.

DEFINING STOCK FOR PART

The **Stock Manager** tool is used to define stock for part. This stock represents the workpiece in actual machining. The procedure to use this tool is given next.

- Click on the **Stock Manager** tool from the **SOLIDWORKS CAM CommandManager** in the **Ribbon**. The **Stock Manager PropertyManager** will be displayed; refer to Figure-11.

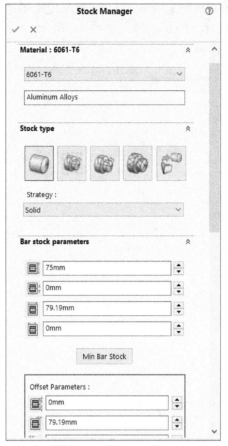

Figure-11. Stock Manager PropertyManager for turning

- By default, the **Round Bar Stock** button is selected in the **Stock type** rollout to define shape of stock. Specify desired size parameters in the **Bar stock parameters** rollout to define size of stock.
- Select the **From Revolved Sketch** button if you want to use revolve feature of selected SolidWorks sketch as stock. Select the **From Revolved 2d WIP File** button if you want to use work in progress file saved after performing other operations as base stock for current setup. Select the **From STL File** button if you want to use STL file for defining stock. Select the **Part File** button if you want to use SolidWorks part file for defining stock.
- Select desired option from the **Strategy** drop-down to define whether your stock is solid or it has hole predrilled to define ID of part. Select the **Cored** option from the drop-down if your part has predrilled hole.
- From the **Material** rollout, set the material of stock. The cutting parameters like feed rate, lead-in, lead-out are defined based on stock material.
- Click on the **OK** button from the dialog box to create the stock.

TURN FEATURE OPTIONS

There are a few settings available in **Options** dialog box to define scope of automatically recognizing turn features. The procedure to define these settings is given next.

- After setting up a turn machine, click on the **SOLIDWORKS CAM Options** button from the expanded **SOLIDWORKS CAM CommandManager** in the **Ribbon**. The **Options** dialog box will be displayed as discussed earlier.

- Click on the **Turn Features** tab in the dialog box. The options will be displayed as shown in Figure-12.

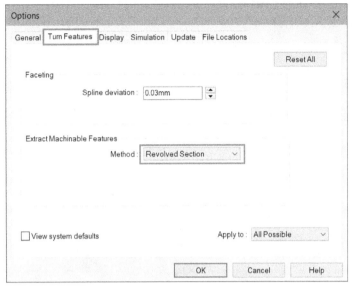

Figure-12. Revolved Section feature extraction method

- Specify the maximum deviation value by which machine model splines can deviate from actual part in the **Spline deviation** edit box.
- Select desired option from the **Method** drop-down of **Extract Machinable Features** area to define how features of model will be recognized for turn machining. There are two methods for recognizing features which are Revolved Section and Plane Section; refer to Figure-13.

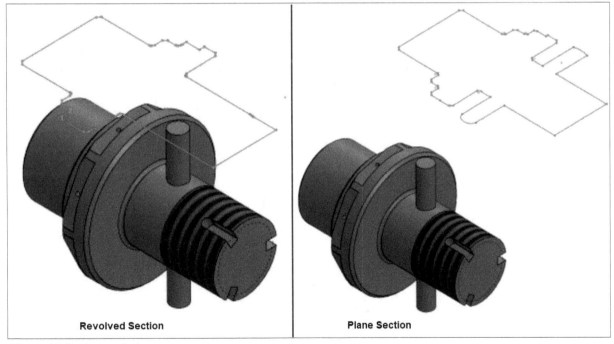

Figure-13. Section method

- Click on the **OK** button to apply settings.

EXTRACTING MACHINABLE FEATURES

The **Extract Machinable Features** tool is used to automatically extract machinable turn features based on stock and part. On clicking this button, the recognized features are displayed in the **SOLIDWORKS CAM NC Tree**; refer to Figure-14.

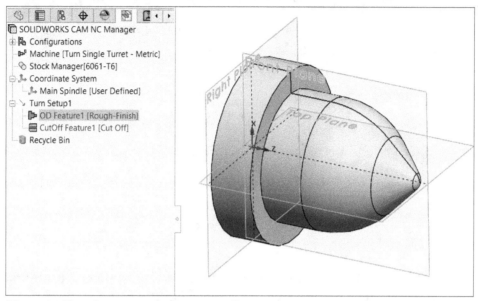

Figure-14. Turn features extracted automatically

CREATING TURN FEATURES MANUALLY

The **Turn Feature** tool is used to manually create CAM features for part. The procedure to use this tool is given next.

- Select the **Turn Setup** option from **SOLIDWORKS CAM Operation Tree** and click on the **Turn Feature** tool from **Tools** > **SOLIDWORKS CAM** > **New** > **Feature** cascading menu; refer to Figure-15. The **New Turn Feature PropertyManager** will be displayed; refer to Figure-16.

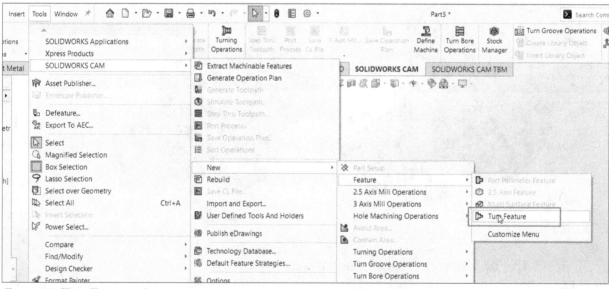

Figure-15. Turn Feature tool

Figure-16. New Turn Feature PropertyManager

- Click on the **Keep visible** button ⇥ if you want to create multiple turn features. On selecting this button, the **PropertyManager** will not exit after creating the feature.
- Select desired option from the **Type** drop-down to define the type of feature to be created for turn machining. The procedures to create different type of turning features are discussed next.

Creating OD Turn Feature

- Select the **OD Feature** option from the **Type** drop-down to machine outer diameter of the part. Note that using this feature, you cannot machine grooves.
- Select desired machining strategy for operation in the **Strategy** drop-down. Select the **Rough-Finish** option if you want to perform roughing and finishing operations for outer diameter of part. Select the **Thread** option if you want to perform threading on the outer face of part. Select the **Rough-Semi finish-Precision** option if you want to rough machine and then finish the part with relatively slower feed for better finish. Select the **Rough w Zero Allowance** option if you want to perform rough machining with better finish.
- Select desired method for creating machining part profile from the **Part Profile Method** drop-down in the **Define from** rollout. These option have been discussed earlier. Note that you can generate part profile at a specified angle if you are using **Plane Section** method. To do so, select the **Plane Section** option from the **Part Profile Method** drop-down and then specify desired angle in the **Turn Section Plane Angle** spinner. After setting the parameters, click on the **Generate Profile** button.
- Click in the selection box of **Selected entities** rollout if you want to add more entities for machining. By default, the automatically generated profile is used for machining but if you select a segment of the profile then only that segment will be machined. Generally, this selection box is kept empty.
- If you want to extend a segment of the automatically generated profile then select it from the model and then specify desired parameters in the **Extend1** and **Extend2** rollouts; refer to Figure-17.

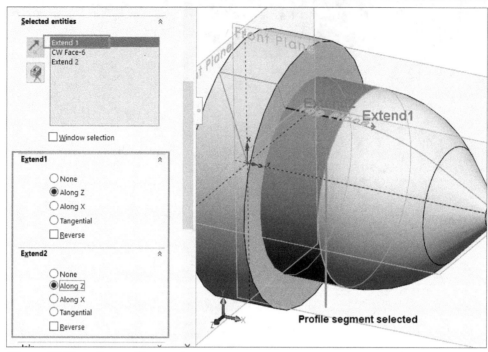

Figure-17. Extending profile segment

- If you have selected segments of profile which have gap between them then a join segment is automatically generated to connect them. For example, in Figure-18 we have selected segment 1 and segment 2. The join segment is generated automatically. Select desired radio button from the **End 1** area of **Join** rollout. Based on selected radio button, the options in **End 2** area will be activated for selection. Note that preview of join segment will be displayed in the graphics area based on selected radio buttons.

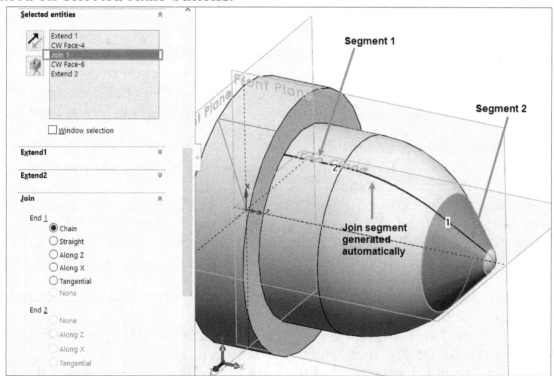

Figure-18. Join segment

- After setting desired parameters, click on the **OK** button from the **PropertyManager**. The feature will be created and displayed in the **SOLIDWORKS CAM Feature Tree**.

Creating ID Feature

The ID feature is used to machine a center hole or undercut in a cylindrical part. The procedure to create this feature is given next.

- Select the **ID Feature** option from the **Type** drop-down in the **Feature** rollout of **New Turn Feature PropertyManager**. The options will be displayed as shown in Figure-19.

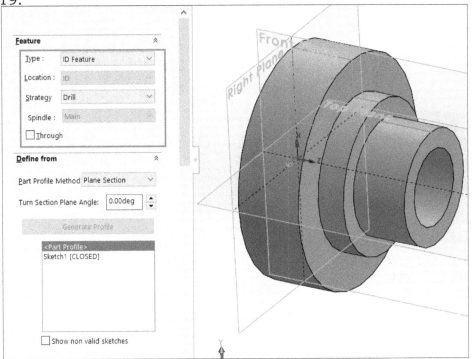

Figure-19. Options for ID Feature

- Select the segment of profile to be used for internal diameter machining.
- Select desired option from the **Strategy** drop-down to define machining strategy. Select the **Drill** option to use drill operation strategy. Select the **Thread** option if you are performing threading in the inner diameter face. Select the **Rough-Finish** option if you want to perform rough machining using drill and then finish by using the boring bar. Select the **Fine** option if you want to perform finishing operation with better finish.
- Select the **Through** check box if hole in part is open at both ends.
- Set the other parameters as discussed earlier and click on the **OK** button from **PropertyManager** to create the feature.

Creating Rectangular Groove

- Select the **Groove Rectangular** option from the **Type** drop-down in the **Feature** rollout of **New Turn Feature PropertyManager**. The options will be displayed as shown in Figure-20.
- Select desired option from the **Location** drop-down to define the location of groove. Select the **OD** option if the groove is on outer diameter face of part as shown in Figure-21. Select the **ID** option if groove is on inner diameter face of part; refer to Figure-21. Select the **Face** option if groove is on front face of model.

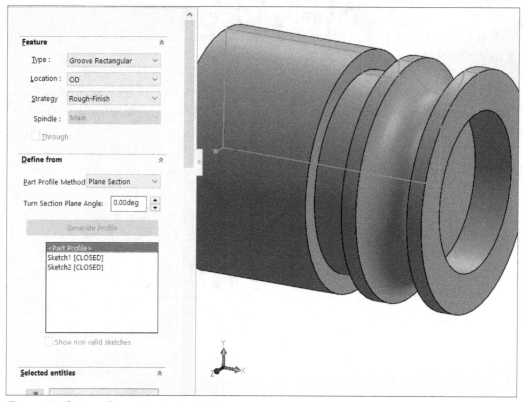

Figure-20. Options for rectangular groove

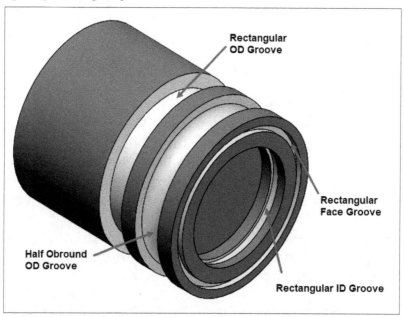

Figure-21. Types of grooves

- If the model has multiple grooves, then click on the **Keep visible** button so that tool does not exit after creating one groove feature.
- After selecting desired option from the **Location** drop-down, select the strategy from the **Strategy** drop-down.
- Select the segment of profile to be machined by groove strategies; refer to Figure-22. Make sure to select both vertical lines of the segment for groove.

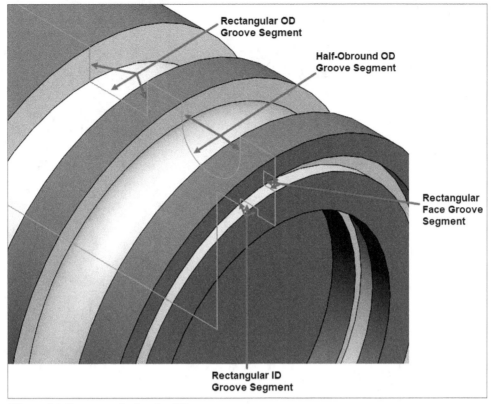

Figure-22. Segment selection for different grooves

- Set the other parameters as discussed earlier and click on the **OK** button.

Creating Half Obround Groove

Half-obround are the half segment of rectangle with semi circular cap on parallel two sides; refer to Figure-23.

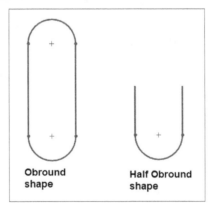

Figure-23. Obround and half-obround shapes

- Select the **Groove HalfObround** option from the **Type** drop-down in the **Feature** rollout of **New Turn Feature PropertyManager**. The options will be displayed accordingly.
- Select the segment for feature; refer to Figure-24. Set the other parameters as discussed earlier and click on the **OK** button.

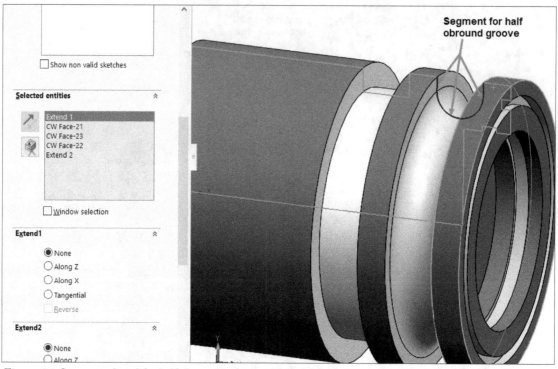

Figure-24. Segment selected for half obround groove

Creating Generic Groove Feature

The **Groove Generic** option in **Type** drop-down of **New Feature PropertyManager** is used to all types of general grooves. The procedure to use this option is same as discussed earlier.

Creating Face Feature

The **Face Feature** option of **Type** drop-down is used to remove material from the front face of model. The procedure to use this option is given next.

- Select the **Face Feature** option from the **Type** drop-down in the **Feature** rollout of **New Turn Feature PropertyManager**. The options will be displayed as shown in Figure-25.
- Select desired strategy for feature from the **Strategy** drop-down for performing machining operations.

Figure-25. Face Feature options

- Select desired segment of profile and define parameters related to join and extend sections if needed; refer to Figure-26.

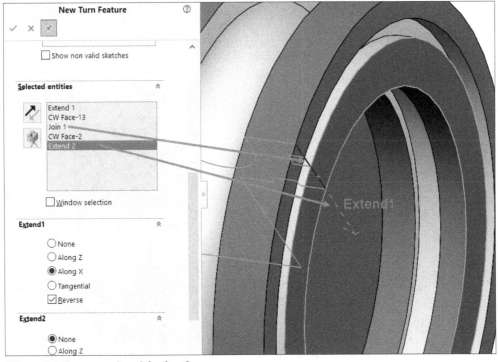

Figure-26. Segment selected for face feature

- Set the other parameters as discussed earlier and click on the **OK** button from the **PropertyManager**. The feature will be created.

Creating CutOff Feature

The Cutoff feature is used to split a part at specified location. The procedure to create this feature is given next.

- Select the **CutOff Feature** option from the **Type** drop-down in the **Feature** rollout of **New Turn Feature PropertyManager**. The options will be displayed as shown in Figure-27.

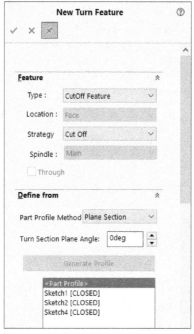

Figure-27. CutOff feature options

- Select the segment of profile to be used for cutoff feature; refer to Figure-28.

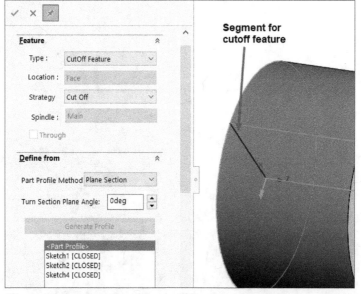

Figure-28. Segment selected for cutoff feature

- Set the other parameters as discussed earlier and click on the **OK** button. The feature will be created.

GENERATING OPERATION PLAN AUTOMATICALLY

The **Generate Operation Plan** tool is used to generate operation plans based on earlier created/extracted turn features. The procedure to use this tool is given next.

- Click on the **Generate Operation Plan** tool from the **SOLIDWORKS CAM CommandManager** in the **Ribbon**. The operations will be created automatically and displayed in the **SOLIDWORKS CAM Feature Tree**; refer to Figure-29.

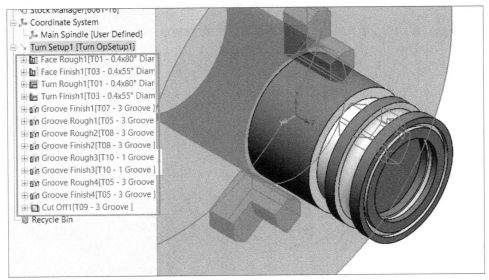

Figure-29. Operations created for turn features

CREATING TURN OPERATIONS MANUALLY

The **Turning Operations** tool is used to create turning operations for various features manually. The procedure to use this tool is given next.

- Click on the **Turning Operations** tool from the **SOLIDWORKS CAM CommandManager** in the **Ribbon**. The **New Operation PropertyManager** will be displayed; refer to Figure-30.

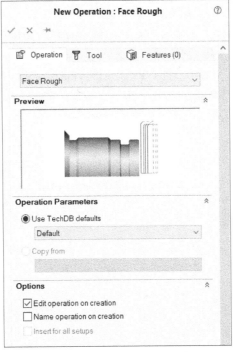

Figure-30. New Operation PropertyManager

Creating Face Rough Operation

- Select the **Face Rough** option from the **Operation Type** drop-down of **Operation** tab in the **PropertyManager**. The options to create face rough operation will be displayed.
- Set desired parameters in the **Operation Parameters** and **Options** rollouts.
- Click on the **Tool** tab and select desired cutting tool from the list; refer to Figure-31.

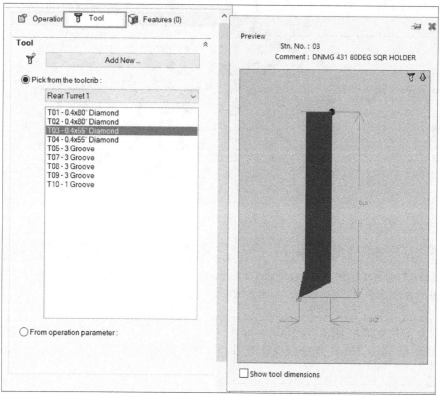

Figure-31. Tool tab for Turning Operations

- Click on the **Features** tab in the **PropertyManager** and select the **Face Feature** check box from the **Pick from the available** area.
- Click on the **OK** button from the **PropertyManager** to create the feature. The **Operation Parameters** dialog box will be displayed with options related to face roughing operation; refer to Figure-32. Note that most of the options in this dialog box have been discussed earlier. So, we will give only overview of earlier discussed options here.

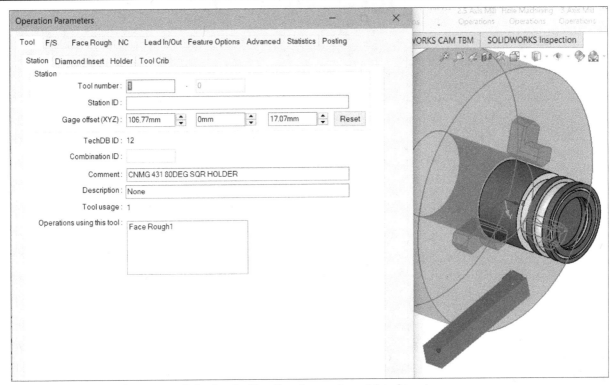

Figure-32. Operation Parameters dialog box for Face Rough operation with tool preview

Tool Tab Parameters

- By default, the **Tool** tab is selected in the dialog box with **Station** sub-tab. Specify desired tool number and station ID in respective edit boxes.
- Set desired values in **Gage offset(XYZ)** edit boxes to offset the position tool holder substation by specified values.
- Specify desired comment, description, and other parameters in the **Station** tab.
- Click on the **Diamond Insert** sub-tab in the dialog box. The options will be displayed as shown in Figure-33.

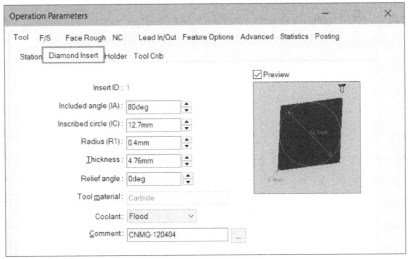

Figure-33. Diamond Insert sub-tab

- The parameters in this tab are standard as per the cutting tool manual. Change these parameters as needed based on tool supplier manual.
- Click on the **Holder** tab and specify the orientation of cutting tool from the **Orientation** area of the dialog box; refer to Figure-34.

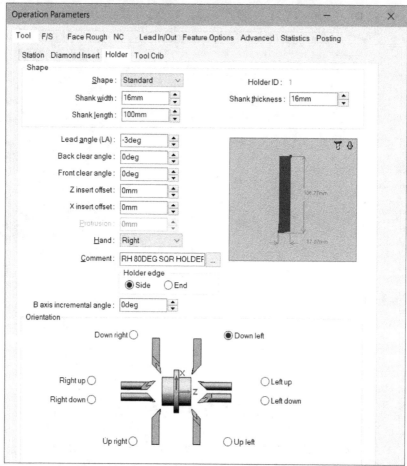

Figure-34. Holder tab of Turning tool

- Specify the parameters related to cutting tool holder like shape, shank width, shank length, shank thickness, clear angles, and so on.
- Select desired option from the **Hand** drop-down to define the cutting edge side of tool holder.
- Select desired radio button from the **Holder edge** area to define reference edge of tool holder. Select the **End** radio button if you want to use end edge of tool holder as reference edge.
- Click on the **Tool Crib** tab to add, delete, or modify order of cutting tools in the tool crib; refer to Figure-35.

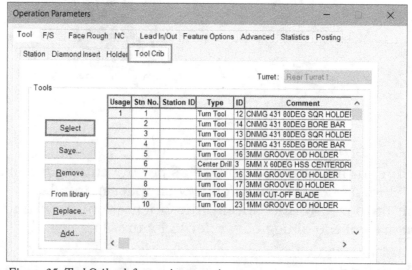

Figure-35. Tool Crib tab for turning operation

Feed and Speed Parameters

- Click on the **F/S** tab in the **Operation Parameters** dialog box. The options will be displayed as shown in Figure-36.

Figure-36. F/S tab for face rough operation

- Select desired option from the **Defined by** drop-down to define whether speed and feed parameters will be based on operation or material library. If you have selected the **Operation** option from the **Define by** drop-down then the options in **Spindle** and **Feed** area will be active.
- Select desired option from the **Mode** drop-down to define whether you want to specify spindle speed in Surface Meter per Minute(SMM) or Round per Minute(RPM).
- Specify desired spindle speed in edit boxes of **Spindle** area.
- Select the **FPM** or **FPR** radio button from the Feed area to define feed rate per minute or feed rater per round, respectively.
- Select the **System Calculated Arc Feedrates** check box to automatically set the feed rates for machining arcs. For inside arc machining, Fi = Feedrate*R/(R+r) where R is external radius and r is internal radius of arc faces. For outside arc machining, Fo = Feedrate*R/(R-r) where R is external radius and r is internal radius of arc faces.
- Select desired check boxes from the **Overrides** area to override respective feedrates. Select the **%** check box if you want to specify override feedrate in percentage.

Face Rough Parameters

- Click on the **Face Rough** tab to display parameters related face roughing operation; refer to Figure-37.

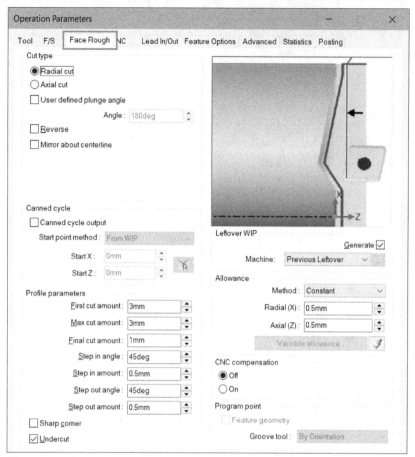

Figure-37. Face Rough tab

- Select the **Radial cut** radio button if you want to perform face cutting operation by step by step decreasing the radius of cut while moving towards workpiece center and then move to next cut depth.
- Select the **Axial cut** radio button to machine the workpiece up to specified depth for current radius then move to closer radius till workpiece is machined up to the specified depth; refer to Figure-38.

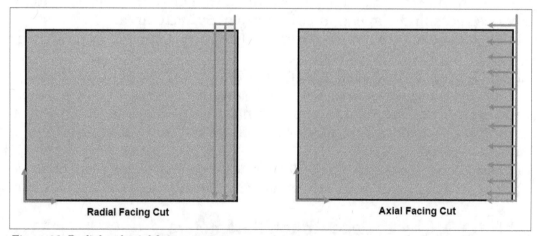

Figure-38. Radial and axial facing cuts

- Select the **User defined plunge angle** check box to define angle at which tool will enter while performing facing operation. Note that the angle at which cut will be performed is perpendicular to plunge angle. After selecting the check box, specify desired angle value in the **Angle** edit box.

- Select the **Reverse** check box to reverse the direction of facing cut. By default, cutting tool moves from outer edge of part to center of the workpiece.
- Select the **Mirror about centerline** check box to reverse the side from where cutting tool will start cutting.
- Select the **Canned cycle output** check box if your CNC machine controllers support canned cycle codes for turning operations like G40, G42, G71, and so on; refer to Figure-39. On selecting this check box, the options in **Canned cycle** area will be activated.

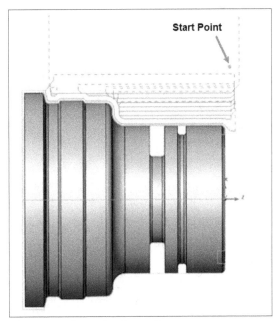

Figure-39. Canned cycle preview

- Select the **Feature Extents** option from the **Start point method** drop-down if you want to use boundary of feature as starting location for operation. Select the **From WIP** option from the **Start point method** drop-down to start facing operation from the end point of previous operation. Select the **User Defined** option from the drop-down if you want to manually specify the starting point of canned cycle. After selecting the **User Defined** option, specify the parameters in the **Start X** and **Start Z** edit boxes.
- Specify desired value in the **First cut amount** edit box to specify the depth of first cut.
- Specify desired value in the **Final cut amount** edit box specify the depth of last cut.
- Specify the maximum amount of depth for general cuts in the **Max cut amount** edit box.
- If the **Canned cycle output** check box is not selected then you can specify angles and amounts for stepping in and stepping out of the workpiece while cutting in respective edit boxes of **Profile parameters** area.
- Select the **Undercut** check box if there is an undercut in the profile with back angle. On selecting this check box, tool will follow entire toolpath including back angles.
- Select the **Sharp corner** check box if you want to make sharp cuts at intersections.
- Select the **Generate** check box from the **Leftover WIP** area if you want to generate work in progress stock for performing current operation.
- Select desired option from the **Machine** drop-down to define stock for current operation. Select the **Initial Stock** option from the drop-down if you want to use original stock and part dimensions for generating toolpath. Select the **Previous**

Leftover option from the drop-down if you want to use stock left after previous operation as current stock. Select the **From Simulation** option if you want to use stock left after running an operation in simulation. After selecting this option, click on the Browse button next to the drop-down. The **Operations for WIP** dialog box will be displayed; refer to Figure-40. Set the other parameters as desired and click on the **OK** button. The stock will be calculated automatically. Select the **New Stock** option if you want to use a new stock for current operation.

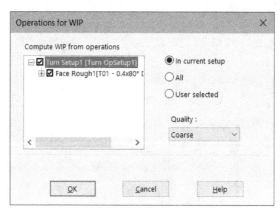

Figure-40. Operations for WIP dialog box2

- Specify desired option from the **Method** drop-down to define method for specifying machining allowance. Select the **Constant** option from the drop-down if you want to specify fixed tolerance in X and Z directions. The allowance value specified here is the maximum allowed deviation of tool from toolpath in respective direction. Select the **Variable** option from the drop-down if tolerance varies for each segment of the profile. After selecting the **Variable** option, click on the **Variable allowance** button. The **Allowance** dialog box will be displayed; refer to Figure-41. Set desired values of radial and axial tolerances for each segment, and click on the **OK** button.

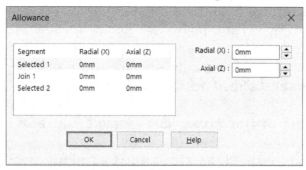

Figure-41. Allowance dialog box

- Select the **On** option from the **CNC compensation** area to specify machine compensation using G41, G42 codes.
- Select the **Feature geometry** check box to use tip of turning tool as program point for performing machining operation. The effect of selecting and clearing the check is shown in Figure-42.
- If you are using a groove tool for facing then select desired program point for the groove tool in **Groove tool** drop-down.

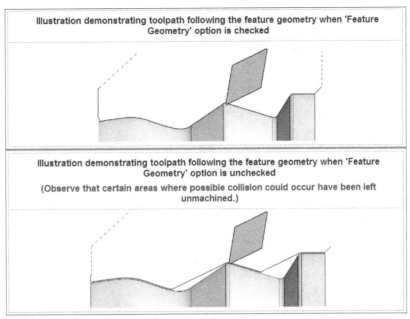

Figure-42. Feature Geometry check box effects

NC Parameters

The options in **NC** tab are used to specify parameters related to clearance, retraction, and approach; refer to Figure-43.

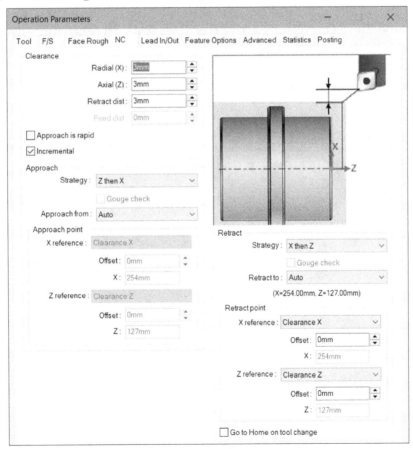

Figure-43. NC tab

- Specify desired value for clearance in the **Radial (X)** and **Axial (Z)** edit boxes. The cutting tool will move away from the current toolpath by specified clearance value after performing a cutting pass.

- Specify desired value in **Retract dist** edit box to define distance by which cutting tool will retract for moving to beginning of next cutting pass.
- Select the **Approach is rapid** check box so that tool approaches at rapid feed rate for making first cut.
- Select the **Incremental** check box to specify clearance distance in incremental coordinates. Otherwise, absolute coordinate will be used to specify clearance distance.
- Select desired option from the **Strategy** drop-down to define how tool will move when approach for start cut. Select the **Z then X** option from the drop-down if you want the cutting tool to first move along Z axis and then move along X axis to perform first cut. Select the **X then Z** option from the drop-down if you want cutting tool to first move along X axis and then Z axis for first cut. Select the **Direction** option from the drop-down if you want to move cutting tool directly to start point of first cut. The tool will move in X and Z axes simultaneously.
- Select the **Gouge check** check box so that entire part shape and collision is avoided during machining.
- Select desired option from the **Approach from** drop-down to define location from where machining will start. Select the **Previous toolpath retract point** option from the drop-down to use retraction point of previous toolpath as start point of current operation. Select the **Approach Point** option from the drop-down to use approach point specified for operation. After selecting this option, the parameters in Approach point area will be active. Specify the parameters as desired.
- You can specify the parameters for retraction and retraction point in the **Retract** area as discussed for clearance.
- Select the **Go to Home on tool change** check box to move cutting tool at home position each time the tool changes.

Lead In/Out

The options in the **Lead In/Out** tab are used to define how toolpaths start and end; refer to Figure-44. These options are discussed next.

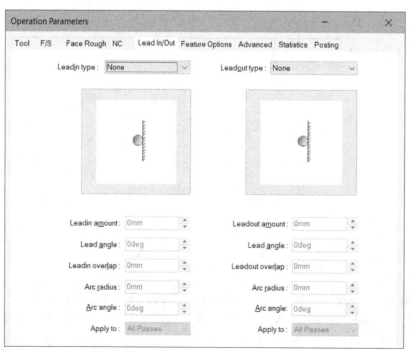

Figure-44. Lead In/Out tab

- Select desired option from the **Leadin type** drop-down to define how toolpath will start. Select the **Arc** option from the drop-down if you want the toolpath to start with an arc. Select the **Perpendicular** option from the drop-down if you want the toolpath to start perpendicular direction of first cut. Select the **Parallel** option from the drop-down to start toolpath parallel to first cut. Select the **None** option from the drop-down if you do not want additional section added in toolpath for lead in.

- Specify desired value in **Leadin amount** edit box to define the distance from where cutting tool will start approach with machining feedrate.

- Specify desired value in the **Lead angle** edit box to define angle with respect to X axis at which tool will approach the workpiece.

- Specify desired value in **Leadin overlap** edit box to define the length from where cutting will start before actual cutting toolpath.

- If you have selected arc type lead in then you can specify arc radius and arc angle in respective edit boxes.

- Select desired option from the **Apply to** drop-down to define the scope of toolpaths to which lead in will be applied. Select the **All Passes** option from the drop-down if you want to apply lead in to all the cutting passes. Select the **First Pass** option to apply lead in to first cutting pass. Similarly, select **Last Pass** or **First And Last** option to apply lead in, respectively.

- If you want to specify same lead out parameters as specified for lead in then select the **Same as leadin** option from the **Leadout type** drop-down.

Feature Options

The options in **Feature Options** tab are used to modify parameters of CAM features selected for the operation; refer to Figure-45. These options are discussed next.

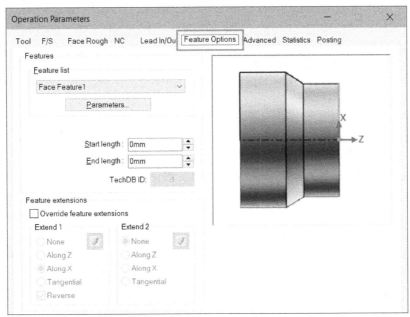

Figure-45. Feature Options tab

- Select desired feature that you want to modify from the **Feature list** drop-down. After selecting the feature, click on the **Parameters** button. The **Face Feature Parameters** dialog box will be displayed; refer to Figure-46.

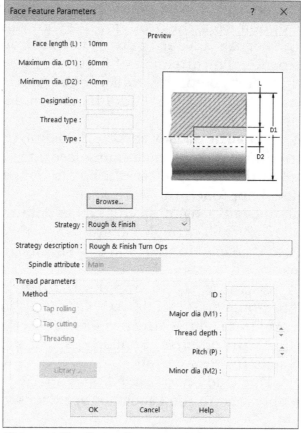

Figure-46. Face Feature Parameters dialog box

- Select desired option from the **Strategy** drop-down to define machining strategy. Select the **Rough & Finish** option to perform roughing and finishing operation. Select the **Thread** option if you want to perform threading on face. After selecting the **Thread** option, specify the thread depth and pitch in respective edit boxes of **Thread parameters** area.

- After setting desired parameters in the dialog box, click on the **OK** button from the **Face Feature Parameters** dialog box. The **Operation Parameters** dialog box will be displayed again.

- Specify desired value in **Start length** edit box to extend starting length of facing toolpath. Similarly, specify the extension length for end toolpath in **End length** edit box.

- Select the **Override feature extensions** check box if you want to extend the length of features as discussed earlier.

Advanced Tab Parameters

- Select the **WIP** option from the **To** drop-down in **Z end** area to use Z depth limit of work piece in progress. If you want to specify the depth value manually, select the **User Defined** option from the **To** drop-down and click on the **Select Point** button next to **Z end** edit box. The **Define Point** dialog box will be displayed; refer to Figure-47.

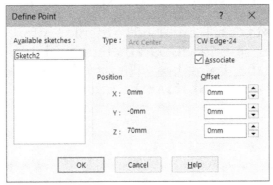

Figure-47. Define Point dialog box

- Select desired reference from the model to define the point; refer to Figure-48.

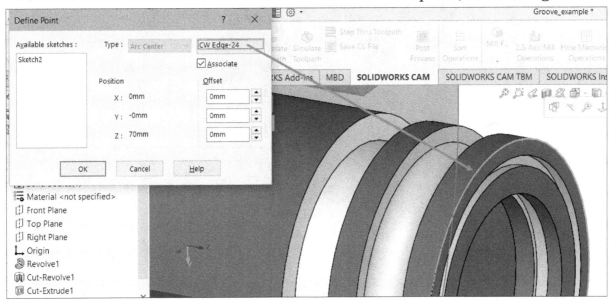

Figure-48. Reference edge selected for point

- After setting desired parameters, click on the **OK** button.
- Select the **Use setup definition** check box if you want to use Z limits based on parameters specified in turn machine setup.
- Specify the parameters in **Z start** area in the same way as discussed for Z end to define position of toolpath in Z direction.
- Set desired options in the **Spline output** area to define the deviation limit of curved toolpath from original geometry. If you want to use deviation value specified in **Options** dialog box for spline toolpaths then select the **Use global spline deviation** check box.
- Select the **Arc fit** check box if you want to replace segments of spline with corresponding arc geometries.
- Select the **Enable** check box from the **Chuck/Fixture avoidance** area to create toolpaths avoiding chucks and fixtures. After selecting the check box, specify desired value in **Clearance** edit box to define distance by which chucks and fixtures will be avoided.
- Select the **Use setup definition** check box to use parameters specified in turn setup.

Statistics Tab

The options in the **Statistics** tab are used to display information related to toolpath and machining time. After setting the parameters in dialog box, click on the **Preview** button at the bottom in the dialog box. The preview of toolpath will be displayed; refer to Figure-49. Click on the **Statistics** tab again to display information; refer to Figure-50.

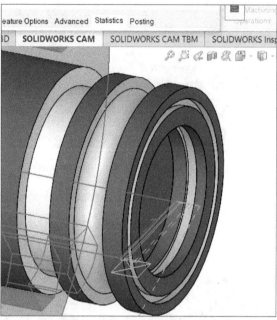

Figure-49. Preview of toolpath

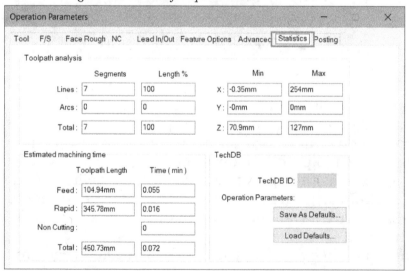

Figure-50. Toolpath information

Posting Parameters

The options in the **Posting** tab are used to define parameters related to coolant, program point, and coordinate system for machining. These options are discussed next.

- Select desired option from the **Absolute Incremental** drop-down whether you want to use absolute coordinates for toolpath programming or incremental values for defining programming coordinates.

- Select desired option from the **Coolant** drop-down to turn on or turn off coolant system during machining.
- Select desired option from the **Program Point** drop-down to define tool nose point to be used as cutting tool point for machining. Select the **Tool Nose Tip** option from the drop-down if you want to use tip of tool as reference for generating toolpaths. Select the **Tool Nose Center** option from the drop-down if you want to use center of cutting tool as reference for generating toolpaths.
- After setting desired parameters, click on the **OK** button. The operation will be created.

Creating Face Finish Operation

The Face finish operation is used to create finishing toolpath for facing. The procedure to create face finish operation is given next.

- Select the **Face Finish** option from the **Operation Type** drop-down in the **Operation** tab of **New Operation PropertyManager**. The options will be displayed as shown in Figure-51.

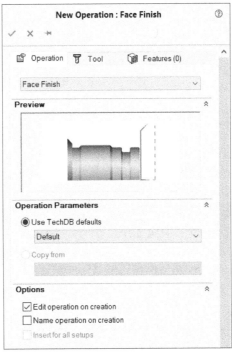

Figure-51. New Operation : Face Finish PropertyManager

- Click on the **Tool** tab and select desired cutting tool for finishing operation.
- Click on the **Features** tab and select desired face feature for finishing operation.
- Click on the **OK** button from the **PropertyManager**. The **Operation Parameters** dialog box will be displayed with face finish options; refer to Figure-52.

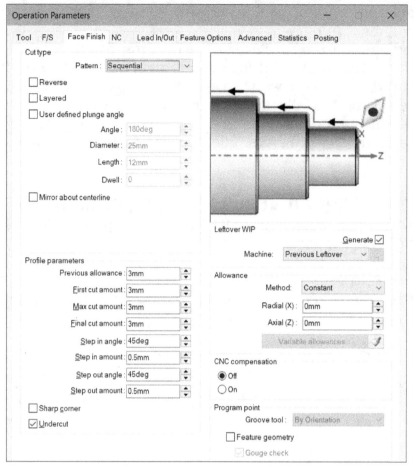

Figure-52. Operation Parameters dialog box for Face Finish operation

- Most of the options in this dialog box are same as discussed earlier for Face rough operation. The options in **Face Finish** tab are discussed next.

Face Finish Options

- Select desired option from the **Pattern** drop-down to define how face finishing operation will be performed. Select the **Sequential** option from the drop-down if you want cutting tool to remove material along full length of feature in each step. Select the **Face down first** option from the drop-down if you want cutting tool to move at the top of face and then go downward while cutting. Select the **Face up first** option from the drop-down if you want cutting tool to move at the bottom of face and then go upward while cutting. Select the **Turn first** option from the drop-down if you want cutting tool to perform turning operation before facing. Select the **Diameter & Length** option from the **Pattern** drop-down if you want cutting tool to reach at desired diameter and length of part for single machining cut and then retract back to home position. Note that the **Diameter & Length** option is generally used for knurling, marking, or o-ring groove machining.
- If you have selected the **Sequential** option from the drop-down then you can reverse the cutting direction by selecting the **Reverse** check box.
- Select the **Layered** check box to perform facing in multiple steps at offset distance equal to **Max cut amount** value specified in this dialog box; refer to Figure-53.

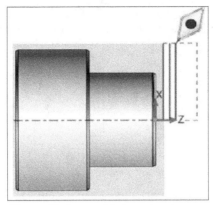

Figure-53. Layered facing operation

- Select the **User defined plunge angle** check box to specify angle at which cutting tool will plunge into the workpiece.
- You can define the other parameters as discussed earlier. Click on the **OK** button from the dialog box to create the face finishing operation.

Creating Turn Rough Operation

- Select the **Turn Rough** option from the **Operation Type** drop-down in the **Operation** tab of **PropertyManager** if you want to create rough turning operation. The **New Operation : Turn Rough PropertyManager** will be displayed; refer to Figure-54.

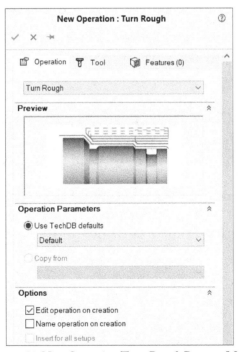

Figure-54. New Operation Turn Rough PropertyManager

- Click on the **Tool** tab and select desired cutting tool for turning operation.
- Click on the **Features** tab and select the OD, groove, or ID features to create operation.
- After setting desired parameters, click on the **OK** button from **PropertyManager**. The options for rough turning will be displayed; refer to Figure-55.

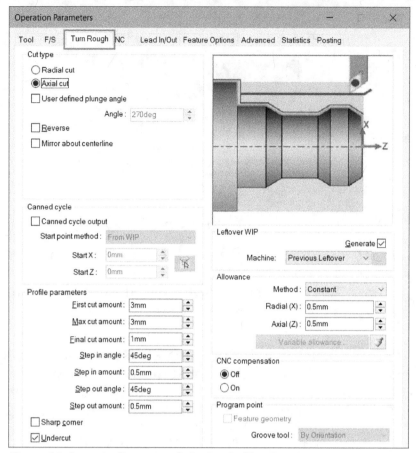

Figure-55. Operation Parameters dialog box for Turn Rough operation

*   Set the parameters as discussed earlier and click on the **OK** button from the dialog box to create the operation.

Creating Turn Finish Operation

The Turn Finish operation is used to remove material from the part after roughing operation to get final part. The procedure to create turn finish operation is given next.

*   Select the **Turn Finish** option from the **Operation Type** drop-down in **Operation** tab of **PropertyManager**. The **New Operation : Turn Finish PropertyManager** will be displayed; refer to Figure-56.

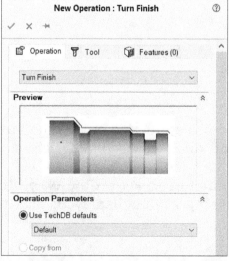

Figure-56. New Operation Turn Finish PropertyManager

- Select desired cutting tool and features for performing turn finish operation. Select the OD, groove, or ID features for which you have earlier created turn rough operations.
- After setting desired parameters, click on the **OK** button from the **PropertyManager**. The **Operation Parameters** dialog box will be displayed with options related to turn finish machining; refer to Figure-57.

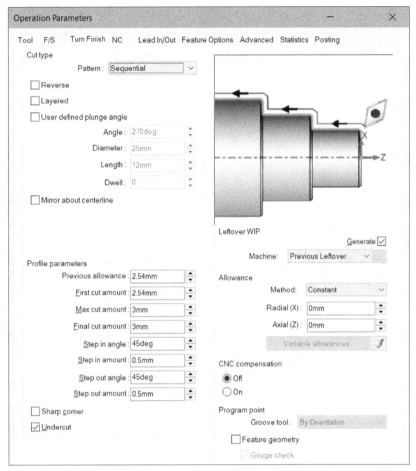

Figure-57. Operation Parameters dialog box for Turn Finish operation

- The options in the dialog box are same as discussed earlier. Set desired parameters and click on the **OK** button. The turn finish operation will be created.

Creating Thread Operation

The thread operation is used to create thread on OD or ID of the part. The procedure to create thread is given next.

- Select the **Thread** option from the **Operation Type** drop-down in the **Operation** tab of **New Operation PropertyManager**. The **New Operation : Thread PropertyManager** will be displayed; refer to Figure-58.

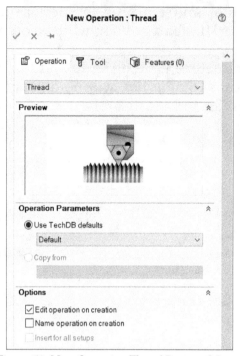

Figure-58. New Operation Thread PropertyManager

- Click on the **Tool** tab and select desired groove or thread cutting tool.
- Click on the **Features** tab and select desired feature on which you want to create threads.
- After setting desired parameters, click on the **OK** button from the **PropertyManager**. The **Operation Parameters** dialog box will be displayed with options related to threading; refer to Figure-59.

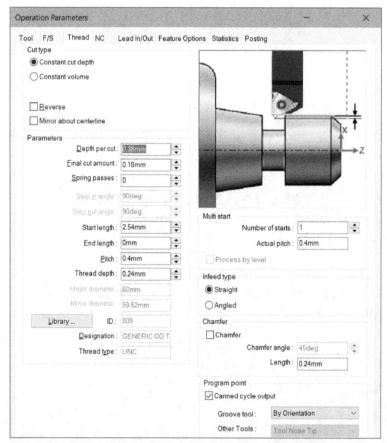

Figure-59. Operation Parameters dialog box for Thread operation

- Select the **Constant cut depth** radio button if you want to keep depth for all the cuts same as specified in **Depth per cut** edit box except for final cut. Select the **Constant volume** radio button from the **Cut type** area if you want the amount of material to be removed in every cut same.
- Select the **Reverse** check box if you want to reverse the direction of cut. By default, cutting tool moves from front face of stock to backward.
- Select the **Mirror about centerline** check box if you want to reverse tool position from the top side of part to bottom side.
- Specify desired value in the **Depth per cut** edit box to define the depth by which cutting tool will move downward after each cutting pass.
- Specify desired value in the **Final cut amount** edit box to define depth for final cut in thread machining.
- Specify desired value in the **Spring passes** edit box to define number of times the final cut will be performed to get good finish of thread.
- Specify desired value in the **Start length** edit box to define distance from original thread where the threading tool will start cutting.
- Click in the **End length** edit box and specify the distance by which threading toolpath will be extended at its end side.
- Specify the distance between two threads in the **Pitch** edit box.
- Specify the depth of thread in the **Thread depth** edit box.
- Click on the **Library** button to select standard thread type for machining. The **Tools Database - Thread Condition (metric)** dialog box will be displayed; refer to Figure-60. Select desired thread and click on the **OK** button.

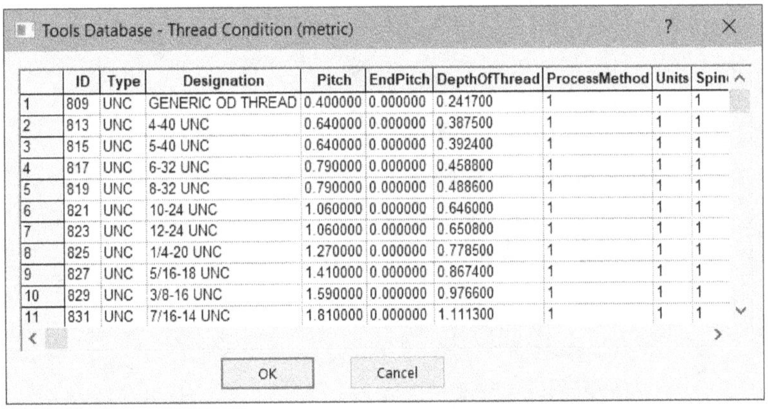

	ID	Type	Designation	Pitch	EndPitch	DepthOfThread	ProcessMethod	Units	Spin
1	809	UNC	GENERIC OD THREAD	0.400000	0.000000	0.241700	1	1	1
2	813	UNC	4-40 UNC	0.640000	0.000000	0.387500	1	1	1
3	815	UNC	5-40 UNC	0.640000	0.000000	0.392400	1	1	1
4	817	UNC	6-32 UNC	0.790000	0.000000	0.458800	1	1	1
5	819	UNC	8-32 UNC	0.790000	0.000000	0.488600	1	1	1
6	821	UNC	10-24 UNC	1.060000	0.000000	0.646000	1	1	1
7	823	UNC	12-24 UNC	1.060000	0.000000	0.650800	1	1	1
8	825	UNC	1/4-20 UNC	1.270000	0.000000	0.778500	1	1	1
9	827	UNC	5/16-18 UNC	1.410000	0.000000	0.867400	1	1	1
10	829	UNC	3/8-16 UNC	1.590000	0.000000	0.976600	1	1	1
11	831	UNC	7/16-14 UNC	1.810000	0.000000	1.111300	1	1	1

Figure-60. Tools Database –Thread Condition dialog box

- If you want to create multi start thread then specify number of threads to be created in the **Number of starts** edit box of the **Multi start** area of the dialog box. Note that based on specified number of starts, the pitch will change in the **Actual pitch** edit box.
- Select the **Process by level** check box if you want to generate all starts at the first depth level and then processes all starts at the next level, and so on for each depth.
- By default, **Straight** radio button is selected in **Infeed type** area so that threads are created in straight line. If you are creating threads in straight line then select the **Angled** radio button from the **Infeed type** area to move cutting tool at an angle equal to back angle of threading tool. A standard thread generally has infeed at 29 degree angle.

- Select the **Chamfer** check box if there is a chamfer in the part. The options to define chamfer angle and length will be displayed. Specify desired values in respective edit boxes based on part geometry.
- Select the **Canned cycle output** check box to use canned cycles for machining threads.
- Set desired option in the **Groove tool** drop-down to define orientation of tool for cutting.
- Set the other parameters as discussed earlier and click on the **OK** button from the dialog box to create threads.

Creating Cut Off Operation

The Cut Off operation is used to split part from the stock at the backside. The procedure to create this operation is given next.

- Select the **Cut Off** option from the **Operation Type** drop-down in the **Operation** tab of **PropertyManager**. The **New Operation : Cut Off PropertyManager** will be displayed; refer to Figure-61.

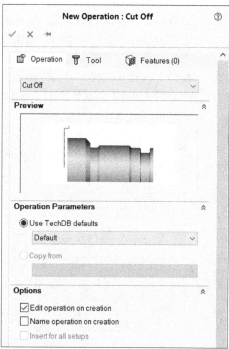

Figure-61. New Operation : Cut Off PropertyManager

- Select desired cutting tool and cut off feature from the **PropertyManager**, and click on the **OK** button. The **Operation Parameters** dialog box will be displayed with cut off parameters; refer to Figure-62.

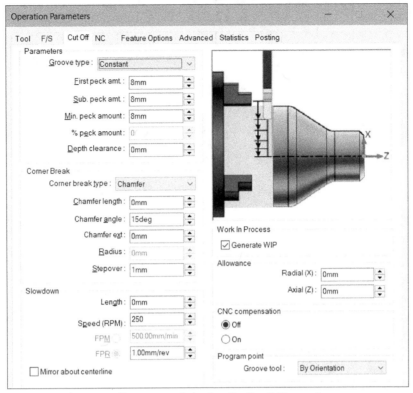

Figure-62. Operation Parameters dialog box for Cut Off operation

- Select desired option from the **Groove type** drop-down to define groove peck distance in percentage or distance value. Select the **None** option if you do not want to use pecking cycles. Select the **Constant** option from the drop-down if you want to specify fixed depth for each pecking cycle. Select the **Percent** option if you want to specify peck depth amount in percentage of Min. peck amount value.

- Specify desired value in the **First peck amt.** edit box to define depth up to which cutting tool will go for first cut.

- Specify desired value in **Sub. peck amt.** edit box to define depth for subsequent cutting pecks after first peck.

- Specify desired value in **Min. peck amt.** edit box to define minimum depth for cutting pecks. Note that even if the material left is less than the minimum peck amount still the value specified in edit box will be used.

- If you have selected **Percent** option in **Groove type** drop-down then specify the value of subsequent cuts in **%peck amount** edit box.

- Specify desired value in **Depth clearance** edit box to define extended depth up to which cutting tool will move past centerline during cut off.

- If you want to create chamfer or radius at the corner after cut off then select the respective option from the **Corner break type** drop-down. Set desired parameters in the **Corner Break** area based on selected option.

- If you want to slow down the feed rate and spindle speed near the center of part then specify the length from center of part at which the cut off machining will slow down in **Length** edit box of the **Slowdown** area.

- Specify desired values in **Speed (RPM)** and **FPR/FPM** edit boxes to define slowed cutting rate.

- Set the other parameters as discussed earlier and click on the **OK** button.

CREATING TURN GROOVE OPERATION

The **Turn Groove Operations** tool is used to machine grooves in the part. The procedure to use this tool is given next.

- Click on the **Groove Rough** tool from the **Turn Groove Operations** drop-down of the **SOLIDWORKS CAM CommandManager** in the **Ribbon**. The **New Operation : Groove Rough PropertyManager** will be displayed; refer to Figure-63.

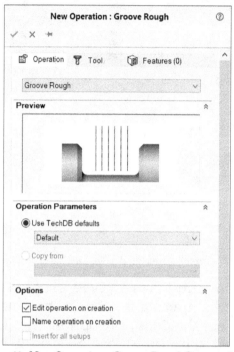

Figure-63. New Operation : Groove Rough PropertyManager

Creating Rough Groove Operation

- Select the **Groove Rough** option from the drop-down to create rough grooving operation.
- Select desired groove cutting tool from the **Tool** tab and select desired groove features to be machined by using the groove turn operation in **Features** tab of **PropertyManager**.
- Click on the **OK** button from the **PropertyManager**. The **Operation Parameters** dialog box will be displayed with groove rough options; refer to Figure-64.
- Specify the options in **Groove peck type** drop-down as discussed earlier.
- Select desired option from the **Groove style** drop-down to define how groove cutting will be performed. Select the **Normal** option if you want the cutting tool to plunge in the form of parallel cuts. Select the **Zig** option from the drop-down if you want the cutting tool to plunge and make cutting moves parallel to bottom face of groove. Select the **Zigzag** option from the drop-down if you want the cutting tool to move zigzag in groove with bidirectional cutting moves; refer to Figure-65.

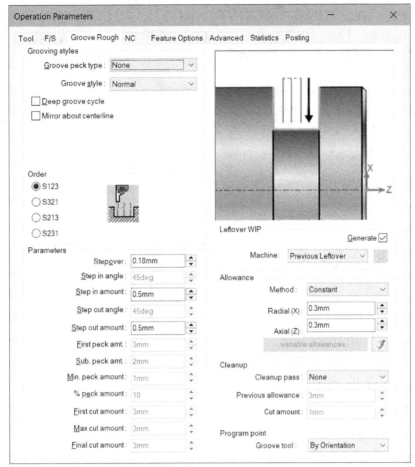

Figure-64. Operation Parameters dialog box for Groove Rough operation

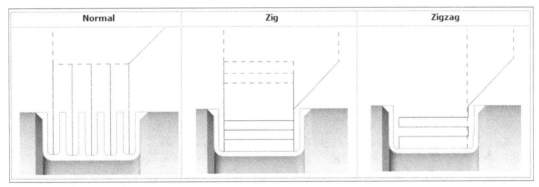

Figure-65. Groove styles

- Select the **Deep groove cycle** check box if you want to cut deep groove in workpiece which require using a thin insert that would break if you could not peck the groove.
- Select the **Mirror about centerline** check box if you want to reverse cutting direction from top to bottom.
- Select desired option from the **Order** area to define order in which cutting tool will plunge into the material.
- Set the other parameters as discussed earlier and click on the **OK** button. The groove rough operation will be created.

Creating Groove Finish Operation

- Select the **Groove Finish** option from the **Operation Type** drop-down in the **Operation** tab of **PropertyManager**. The related options will be displayed in **PropertyManager**.

- Set desired parameters in the **PropertyManager** and click on the **OK** button. The **Operation Parameters** dialog box will be displayed with groove finish options; refer to Figure-66.

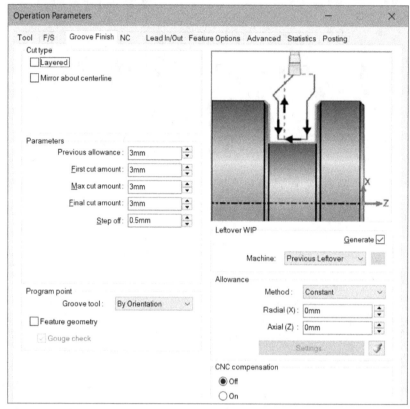

Figure-66. Operation Parameters dialog box for Groove Finish operation

- Set the parameters in dialog box as discussed earlier and click on the **OK** button. The groove finish operation will be created.

CREATING TURN BORE OPERATIONS

The **Turn Bore Operations** tool is used to create operations like bore rough, bore finish, drill, central drill, tap and threading. The procedure to use this tool is given next.

- Click on the **Turn Bore Operations** tool from the **SOLIDWORKS CAM CommandManager** in the **Ribbon**. The **New Operation PropertyManager** will be displayed; refer to Figure-67.

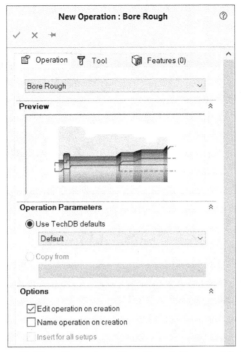

Figure-67. New Operation : Bore Rough PropertyManager

Creating Bore Rough Operation

The Bore Rough operation is performed to remove large amount of material from the hole to increase its diameter which cannot be achieved by drill bit.

- Select the **Bore Rough** option from the **Operation Type** drop-down in **PropertyManager**. Select desired cutting tool and feature to be machined by bore roughing.
- After setting desired parameters, click on the **OK** button from the **PropertyManager**. The **Operation Parameters** dialog box will be displayed with bore rough options; refer to Figure-68.
- The options in this dialog box are same as discussed earlier. After setting desired parameters, click on the **OK** button from the dialog box.

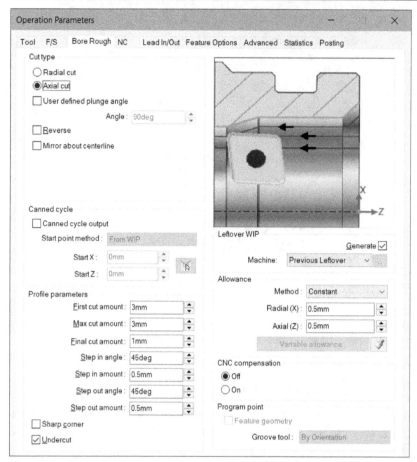

Figure-68. Operation Parameters dialog box for Bore Rough operation

Creating Bore Finish Operation

The Bore Finish operation is performed to get desired finish in bore after roughing operation. The procedure to create bore finish operation is given next.

- Select the **Bore Finish** option from the **Operation Type** drop-down in the **Operation** tab of **PropertyManager**. Select desired features and cutting tool for the operation, and click on the **OK** button. The **Operation Parameters** dialog box will be displayed with bore finish options; refer to Figure-69.

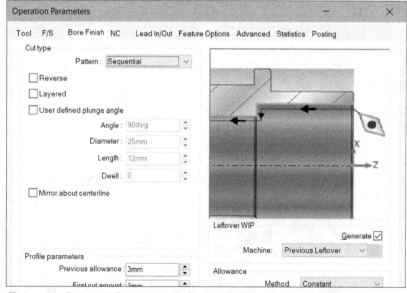

Figure-69. Operation Parameters dialog box for Bore Finish operation

- Set desired parameters as discussed earlier and click on the **OK** button. The bore finish operation will be created.

Creating Drill Operation

The Drill operation is performed to machine holes using the drill bit cutting tools. The procedure to create drill operation is given next.

- Select the **Drill** option from the **Operation Type** drop-down in the **Operation** tab of **PropertyManager**. Select the drill of desired size from the **Tool** tab and select the feature to be used for drilling from the **Features** tab.
- After setting desired parameters, click on the **OK** button. The **Operation Parameters** dialog box will be displayed with drill parameters; refer to Figure-70.

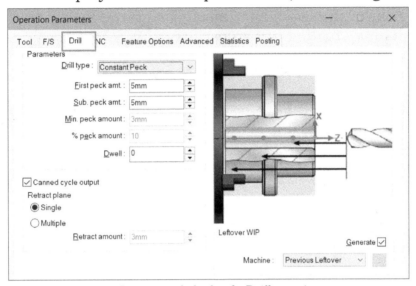

Figure-70. Operation Parameters dialog box for Drill operation

- Select the **Drilling** option from the **Drill type** drop-down if you want to drill hole in one step. The drill will reach till the specified depth in one go. Select the **Constant Peck** option from the drop-down if you want to create peck drilling with constant size of pecks. Select the **Percent Peck** option from the **Drill type** drop-down if you want to specify peck amount in percentage of minimum peck amount.
- Set the other parameters in the dialog box as discussed earlier and click on the **OK** button. The operation will be created.

Similarly, you can create the Center Drill, Tap, and Thread operations using respective options.

You can generate, simulate, export toolpaths as discussed in Chapter 5.

SELF ASSESSMENT

Q1. Using the Define Machine tool, you can create setup for turning machine. (T/F)

Q2. Discuss the use of work offset for turning operations.

Q3. If you have hollow workpiece then you can select the **Jaws In** radio button from **Chuck Parameters PropertyManager** so that jaws hold the workpiece using inner diameter of stock. (T/F)

Q4. There are two methods for recognizing features which are .... and Plane Section.

Q5. If the model has multiple grooves, then click on the Keep visible button so that tool does not exit after creating one groove feature. (T/F)

Q6. What are the full forms of SMM and RPM regarding spindle speed?

Q7. Select the **Gouge check** check box so that entire part shape and collision is avoided during machining. (T/F)

Q8. A standard thread generally has infeed at .... degree angle.

Chapter 8

Practical and Practice

Topics Covered

The major topics covered in this chapter are:

- *Introduction*
- *Practical 1*
- *Practical 2*
- *Practical 3*
- *Practice 1*
- *Practice 2*
- *Practice 3*

INTRODUCTION

Till this chapter, you have worked on various tools related to Milling operations and turning operations using CAM. In this chapter, we will learn to apply the techniques in real examples.

PRACTICAL 1

Create CAM program for part shown in Figure-1. The stock for this part is a rectangular block of 200mm x 120mm x 55mm.

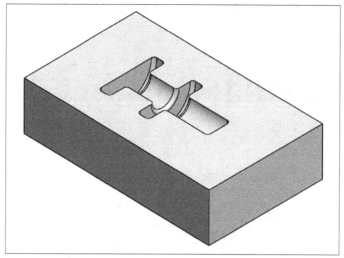

Figure-1. Model for Practical 1

Steps Overview:
1. Start SolidWorks with SolidWorks CAM add-in loaded.
2. Open the part file for Practical 1 of Chapter 8.
3. Define Milling machine and stock.
4. Create the milling setup and extract the milling features.
5. Create the milling features if needed, generate the operations, and generate the toolpaths.
6. Check the simulation and post process.

These steps are discussed next in detail.

Starting SolidWorks and Loading CAM add-in

• Double-click on SolidWorks icon from desktop or **Start** menu. The SolidWorks application will start.
• Click on the **Add-Ins** option from the **Settings** drop-down in the **Quick Access Toolbar**. The **Add-Ins** dialog box will be displayed.
• Select the **SOLIDWORKS CAM** check box from the **Active Add-ins** column in the dialog box. If you want to load **SOLIDWORKS CAM** add-in each time SolidWorks starts then select the corresponding check box from **Start Up** column.
• Click on the **OK** button from the dialog box. The add-in will be loaded.

Loading the File

• Click on the **Open** button from the **Quick Access Toolbar** or press **CTRL+O** from keyboard. The **Open** dialog box will be displayed; refer to Figure-2.

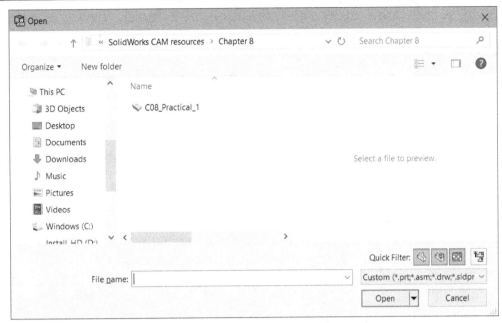

Figure-2. Open dialog box

- Select the part file C08 Practical 1 from the dialog box and click on the **Open** button. The model file will be displayed.

Defining Machine and Creating Stock

- Click on the **Define Machine** tool from the **SOLIDWORKS CAM CommandManager** in the **Ribbon**. The **Machine** dialog box will be displayed; refer to Figure-3.

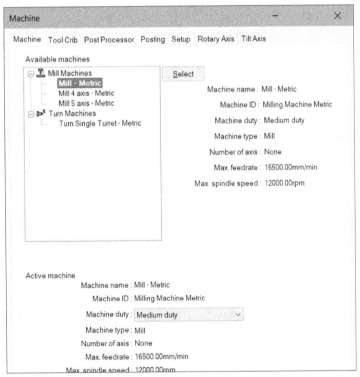

Figure-3. Machine dialog box

- Select the **Mill - Metric** option from the **Mill Machines** node and click on the **Select** button.
- Click on the **Tool Crib** tab in the dialog box to modify available cutting tools. There some important factors for selecting the cutting tool like minimum diameter

for cutting, maximum depth of tool required for cutting, tip angle, and material of cutting tool. You can use the **Measure** tool to measure various important parameters.

- Click on the **Measure** tool from the **Evaluate CommandManager** in the **Ribbon**. The **Measure** toolbox will be displayed; refer to Figure-4.

Figure-4. Measure toolbox

- Measure the distance between two faces which have least distance between them; refer to Figure-5. This will define the minimum diameter of cutting tool. The distance in this case is 12 mm so cutting tool diameter should be less than 12 mm for cutting material. Click on the down arrow button to expand the dialog box. The options will be displayed as shown in Figure-6.

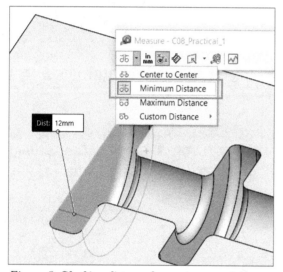

Figure-5. Checking distance for maximum tool diameter

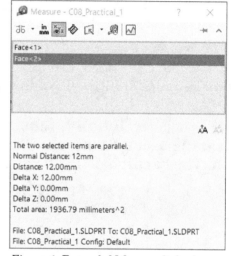

Figure-6. Expanded Measure dialog box

- Right-click in the selection box and select the **Clear Selections** option. Now, select the top face and bottom face of features to determine the length of cutting tool.
- Based on the information collected, make sure you have cutting tools to perform machining. Close the **Measure** box to exit.
- If there is any cutting tool required, click on the **Add Tool** button and select the desired cutting tool.
- Click on the **Post Processor** tab in the dialog box and select the post processor based on machine you have. In this case, select the **MILL\FANUC16M** option from the dialog box.
- Click on the **OK** button from the dialog box to create the machine.
- Click on the **Coordinate System** button from the **SOLIDWORKS CAM CommandManager** in the **Ribbon**. The **Fixture Coordinate System PropertyManager** will be displayed; refer to Figure-7.

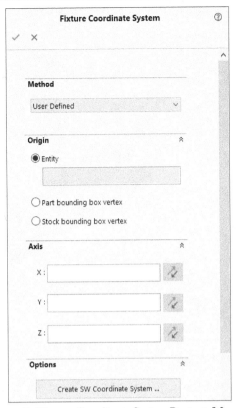

Figure-7. Fixture Coordinate System PropertyManager

- Select the point on the model as shown in Figure-8 and define reference for Z axis.

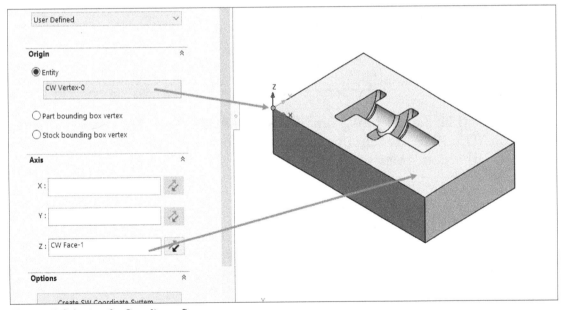

Figure-8. Selection for Coordinate System

- Click on the **OK** button from the **PropertyManager** to define the coordinate system.

Creating Stock

- Click on the **Stock Manager** tool from the **SOLIDWORKS CAM CommandManager** in the **Ribbon**. The **Stock Manager PropertyManager** will be displayed with preview of stock; refer to Figure-9.

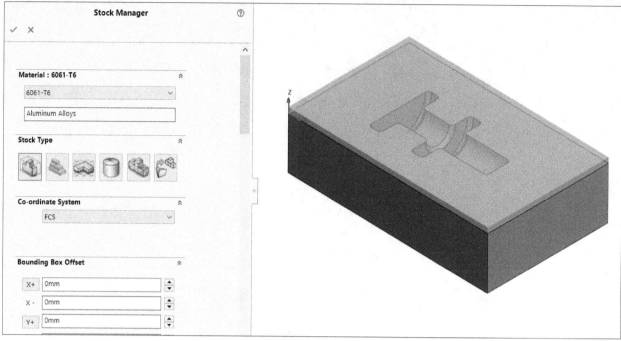

Figure-9. Stock Manager PropertyManager with preview of stock

- Select the **FCS** option from the **Coordinate System** drop-down to use recently created coordinate system.
- Type **5** in the **Z+** edit box to add 5 mm stock material in positive Z direction.
- You can change the material from the **Material** drop-down as needed. We will leave it to default in our case.
- Click on the **OK** button from the **PropertyManager** to create the stock.

Creating Mill Setup

The Mill setup is used to define direction of cutting tool and placement of workpiece.

- Click on the **Mill Setup** tool from the **SOLIDWORKS CAM TBM CommandManager** in the **Ribbon**. The **Mill Setup PropertyManager** will be displayed with preview of machining direction; refer to Figure-10.

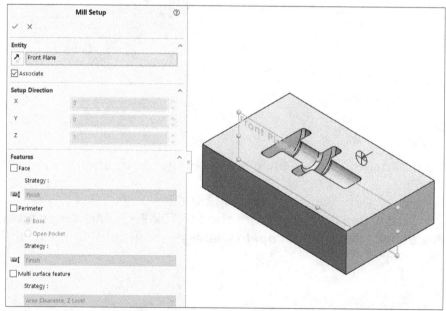

Figure-10. Mill Setup PropertyManager with preview

- Click in the **Entity** selection box and select top face of part to define machining direction; refer to Figure-11.

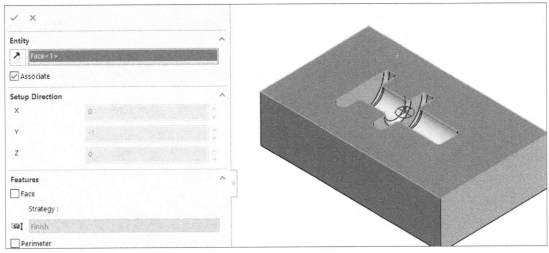

Figure-11. Face selected for mill setup

- Select the **Face** check box from the **Features** rollout to create face feature. Select the **Finish** option from the **Strategy** drop-down.
- Click on the **OK** button from the **PropertyManager** to create the setup.

Creating Machinable Features

- Click on the **Extract Machinable Features** tool from the **SOLIDWORKS CAM CommandManager** in the **Ribbon**. A new mill setup will be created with irregular slot feature; refer to Figure-12.

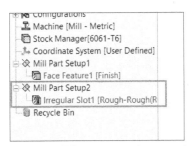

Figure-12. Mill feature created automatically

- Note that this feature and mill setup direction is not as desired for this practical. So, we need to delete this setup. Right-click on the **Mill Part Setup2** option from the **SOLIDWORKS CAM Feature Tree** and select the **Delete** option from the shortcut menu; refer to Figure-13. A message box will be displayed.

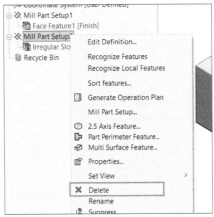

Figure-13. Delete option

- Select the **Yes** option from the dialog box. Right-click on **Recycle Bin** node from the **SOLIDWORKS CAM Feature Tree** and select the **Empty** option to clear recycle bin.
- Select the **Mill Part Setup1** node from the **SOLIDWORKS CAM Feature Tree** and right-click on it. A shortcut menu will be displayed. Select the **Multi Surface Feature** tool from the shortcut menu. The **Multi Surface Feature PropertyManager** will be displayed.
- Select all the faces of pocket that you want to machine; refer to Figure-14.

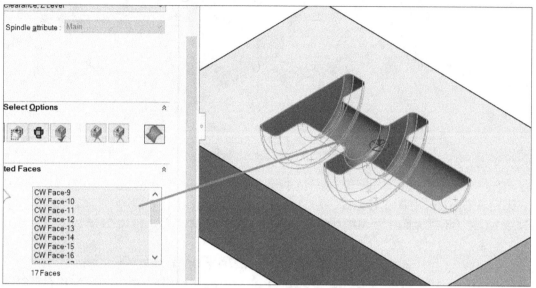

Figure-14. Faces selected for multi surface feature

- Select the **Area Clearance, Z Level** option from the **Strategy** drop-down in the **PropertyManager** and click on the **OK** button.

Creating Operations
Facing Operation

- Click on the **Face Mill** tool from **Tools > SOLIDWORKS CAM > New > 2.5 Axis Mill Operations** menu. The **New Operation PropertyManager** will be displayed.

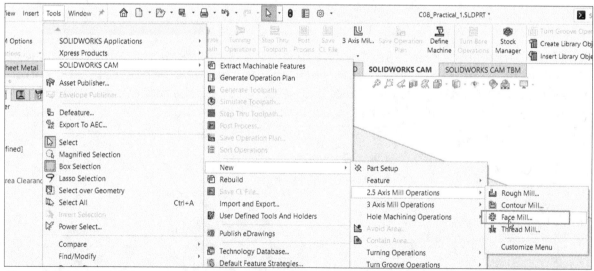

Figure-15. Face Mill tool

- Select the **Edit operation on creation** check box if you want to modify parameters for face mill operation from **Operation** tab in **PropertyManager**. In our case, we are leaving this check box cleared.

- Click on the **Tool** tab in **PropertyManager** and select the **T12 - 50 Face Mill** tool from the list.
- Click on the **Features** tab and select the check box for **Face Feature** in the **Pick from the available** area.
- Click on the **OK** button from the **PropertyManager**. The face mill feature will be created.

Area Clearance Roughing Operation

- Click on the **Area Clearance** tool from **3 Axis Mill Operations** drop-down of the **SOLIDWORKS CAM CommandManager** in the **Ribbon**. The **New Operation PropertyManager** will be displayed.
- Select the **Area Clearance** option from the **Operation Type** drop-down in the **Operation** tab and **Cavity** option from the **Use TechDB defaults** drop-down in the **PropertyManager**.
- Select the **Edit operation on creation** check box to modify operation parameters after exiting the **PropertyManager**.
- Click on the **Tool** tab and click on the **Add New** button. The **Tool Select Filter** dialog box will be displayed.
- Select the **Ball Nose** option from the **Tool type** drop-down and diameter range from 6 mm to 8 mm; refer to Figure-16.

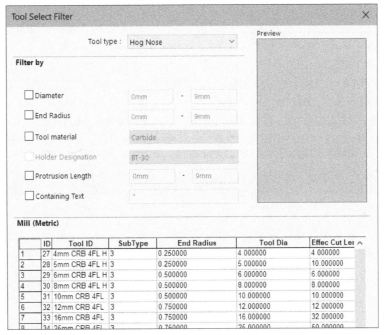

Figure-16. Tool Select Filter dialog box

- Select the tool with ID **43** in the table and click on the **OK** button.
- Click on the **Features** tab and select the check box for **Multi Surface Feature** from the **Pick from the available** area.
- Click on the **OK** button from the **PropertyManager**. The **Operation Parameters** dialog box will be displayed with parameters related to area clearance toolpath; refer to Figure-17.

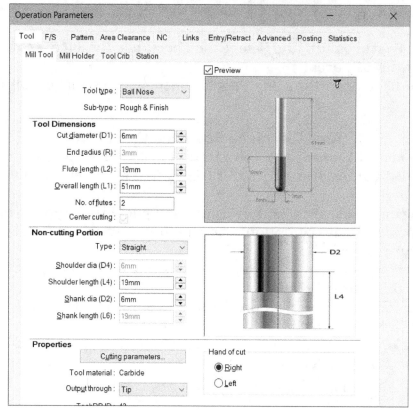

Figure-17. Operation Parameters dialog box for area clearance operation

- Increase the overall length of tool to **70** mm in the **Overall length (L1)** edit box.
- Click on the **Pattern** tab to define the pattern/design in which cutting tool will move in XY plane for cutting.
- Select the **Lace** option from the drop-down to make parallel cuts and specify the distance between two parallel cutting passes as **40%** of tool diameter in the **Lace stepover %** edit box; refer to Figure-18.

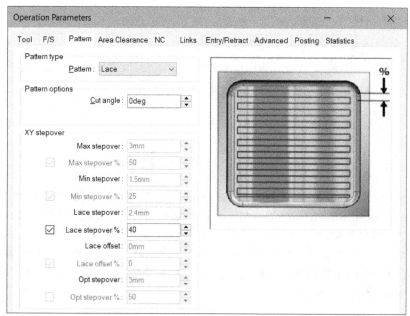

Figure-18. Specifying lace pattern parameters

- Click on the **Area Clearance** tab to define surface finish, depth and other related parameters. Specify the material to be left for finishing toolpath after area clearance machining in **XY Allowance** and **Z allowance** edit boxes. We leave these values to **0.5** mm.
- Select the **Zigzag** check box to make cutting passes in both going and coming back directions of tool.
- Select the cutting method as **Constant** from the **Method** drop-down so that each cutting pass has equal depth of specified value.
- Specify the value of maximum cut depth lower than the cutting length of tool. Note that this value should be less than or equal to half of cutting length of tool to lower load on the tool. In our case, we have specified **13** mm as maximum cut length in the **Max. cut amount** edit box.
- Specify the value of cut depth in Z direction for each cutting pass in **Cut amount** edit box. We have specified the value as **2.5** mm.
- Select the **Adaptive stepdown** check box to create toolpath depths based on parameters specified for maximum cut amount, cut amount, and minimum cut amount. The first cut will be made using maximum cut amount depth, next all cuts will be adaptive to geometry of part. The general cuts will have depth value between cut amount and minimum cut amount values.
- Set the other parameters as shown in Figure-19.

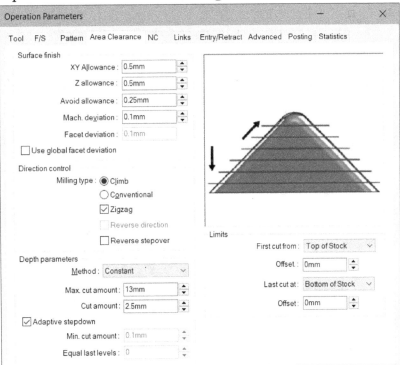

Figure-19. Area clearance parameters specified

- Select the **To depth by region** radio button from the **Depth processing** area in **Links** tab to machine a pocket completely before moving to next pocket.
- Specify desired parameters in the **Corners** area to make round corners of toolpath; refer to Figure-20.

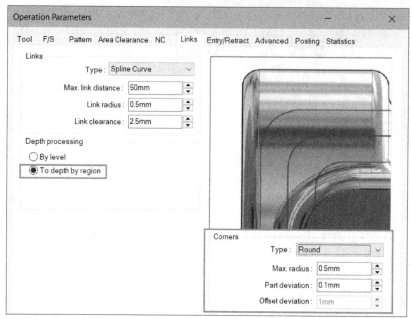

Figure-20. Parameters specified for Links

- Click on the **Advanced** tab to define containment area of features. If containment is not defined properly then your cutting tool may wander away from real stock to be removed.
- Select the **All Silhouettes** option from the **Method** drop-down to use boundaries of surfaces selected for feature as containment area.
- Select the **Upto** option from the **Tool condition** drop-down to keep the cutting edge of tool over containment boundary. The toolpaths will be generated accordingly; refer to Figure-21.

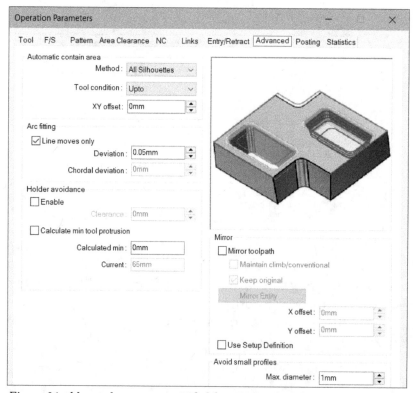

Figure-21. Advanced parameters specified for containment area

- Click on the **Preview** button to check preview of toolpath; refer to Figure-22.

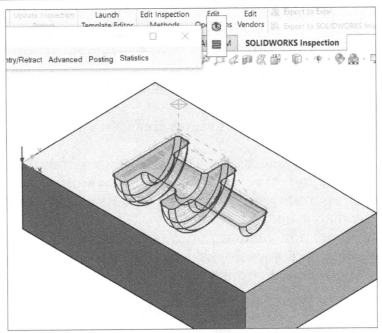

Figure-22. Preview of area clearance toolpath

- Click on the **Simulate Toolpath** button to check how tool will move over toolpath. The **Simulate Toolpath PropertyManager** will be displayed.
- Set the **Pause** option from Tool, Tool Holder, and Tool Shoulder/Shank to check if there is a possible collision during machining; refer to Figure-23.

Figure-23. Setting collision pause parameters

- Click on the **Run** button from **Navigation** rollout to check the simulation.
- Look closely over the areas where you might get problem during machining. After checking the simulation, click on the **OK** button.
- Click on any tab apart from previously selected in the dialog box to display the options of **Operation Parameters** dialog box. If there is a need to modify any parameter then do so in the dialog box. In our case, there is no collision or suspected machining problem.
- Click on the **OK** button to create the operation.

Creating Z Level Finishing Operation

- Click on the **Z Level** tool from **3 Axis Mill Operations** tool from the **SOLIDWORKS CAM CommandManager** in the **Ribbon**. The **New Operation PropertyManager** will be displayed.

- Select the **Default** option from the **Use TechDB defaults** drop-down in Operation tab to use default cutting strategy.

- Select the **Edit operation on creation** check box to modify parameters after exiting the **PropertyManager**.

- Click on the **Tool** tab and select T14 - 6 Ball Nose tool from the **PropertyManager**.

- Click on the **Features** tab and select check box for Multi Surface Feature1.

- After setting desired parameters, click on the **OK** button. The **Operation Parameters** dialog box will be displayed.

- Set the mill tool and mill holder parameters as specified for area clearance operation earlier; refer to Figure-24. Note that your tool should have cutting length and shank length protruded enough from the mill holder to avoid accident.

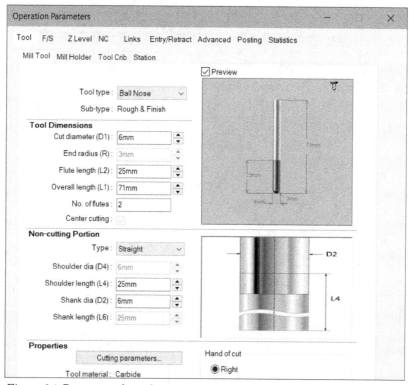

Figure-24. Parameters for tool

- Click on the **Z Level** tab to modify parameters related to Z level toolpath. Specify the allowances as **0** for finishing the part.

- Select the **Zigzag** radio button from **Direction control** area to perform cutting in both directions.

- Set the parameters as shown in Figure-25.

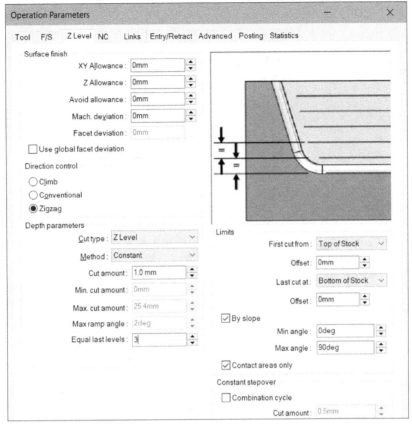

Figure-25. Parameters in Z level tab

- Click on the **Link** tab and set the link type to **Straight Line**. Set the cut depth to 100% max stock and set the depth processing by region; refer to Figure-26.

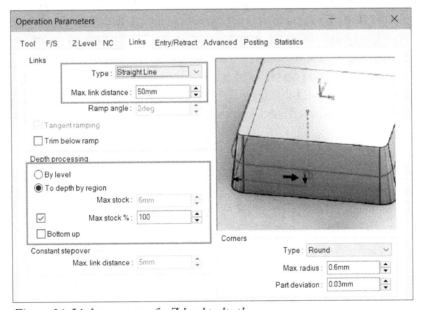

Figure-26. Link parameters for Z level toolpath

- Click on the **Advanced** tab and specify the containment boundaries as shown in Figure-27.

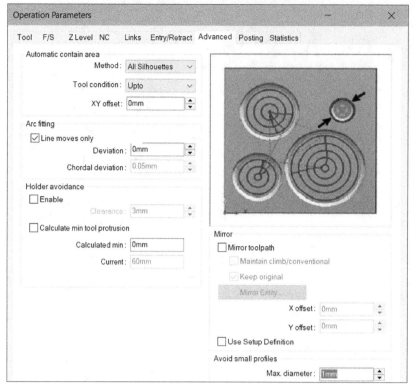

Figure-27. Defining containment area for Z Level operation

- You can check the preview and simulation of toolpath as discussed earlier.
- Click on the **OK** button from the dialog box to create the toolpath.

Generating Toolpaths and Performing Simulation

- Select the **Mill Part Setup** node from the **SOLIDWORKS CAM Operation Tree** and click on the **Generate Toolpath** tool from the **SOLIDWORKS CAM CommandManager** in the **Ribbon**. The toolpath will be created. Close **SOLIDWORKS CAM Process Manager** if displaying.
- Click on the **Simulate Toolpath** tool while the node is selected in the **SOLIDWORKS CAM Operation Tree**. The **Simulate Toolpath PropertyManager** will be displayed.
- Set the parameters as discussed earlier and click on the Run button to simulate toolpath.

Post Processing Toolpaths

- Click on the **Post Process** tool from the **Ribbon**. The **Post Output File** dialog box will be displayed; refer to Figure-28.

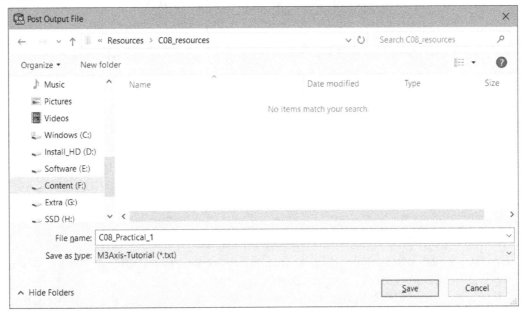

Figure-28. Post Output File dialog box

- Select the desired file format recognized by your machine from **Save as type** drop-down.
- Specify desired name for file in the **File name** edit box and click on the **Save** button. The **Post Process PropertyManager** will be displayed.
- Click on the **Play** button to start generating codes. Once the codes are generated, click on the **OK** button from **PropertyManager**.
- Save the file and close it by pressing **CTRL+W**.

PRACTICAL 2

Create CAM program and generate CNC codes for the model shown in Figure-29. The stock for part is 200mm x 125mm x 40mm.

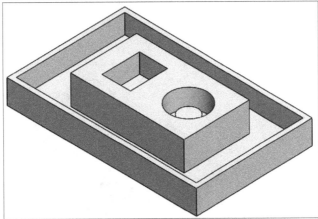

Figure-29. Model for Practical 2

Steps Overview:
1. Start SolidWorks with SolidWorks CAM add-in loaded.
2. Open the part file for Practical 2 of Chapter 8.
3. Define Milling machine and stock.
4. Create the milling setup and extract the milling features.
5. Create the milling features if needed, generate the operations and generate the toolpaths.
6. Check the simulation and post process.

These steps are discussed next in detail.

Starting SolidWorks and Loading CAM add-in

*   Double-click on SolidWorks icon from desktop or **Start** menu. The SolidWorks application will start.
*   Click on the **Add-Ins** option from the **Settings** drop-down in the **Quick Access Toolbar**. The **Add-Ins** dialog box will be displayed.
*   Select the **SOLIDWORKS CAM** check box from the **Active Add-ins** column in the dialog box. If you want to load **SOLIDWORKS CAM** add-in each time SolidWorks starts then select the corresponding check box from **Start Up** column.
*   Click on the **OK** button from the dialog box. The add-in will be loaded.

Loading the File

*   Click on the **Open** button from the **Quick Access Toolbar** or press **CTRL+O** from keyboard. The **Open** dialog box will be displayed.
*   Select the part file C08 Practical 2 from the dialog box and click on the **Open** button. The model file will be displayed.
*   Using the **Measure** tool, check various parameters necessary for defining tool diameter and tool length.

Defining Machine and Creating Stock

*   Click on the **Define Machine** tool from the **SOLIDWORKS CAM CommandManager** in the **Ribbon**. The **Machine** dialog box will be displayed.
*   Set the parameters as discuss in previous practical.
*   Click on the **OK** button to apply the machine parameters.
*   Click on the **Coordinate System** tool from the **Ribbon**. The **Fixture Coordinate System PropertyManager** will be displayed.
*   Click in the **Entity** selection box and select the point to be used for coordinate system as shown in Figure-30. Also, specify the Z direction axis as per the figure.

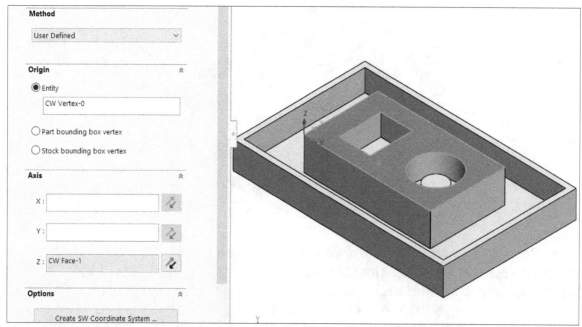

Figure-30. Setting coordinate system

*   Click on the **OK** button from the **PropertyManager** to create the coordinate system.

Creating Stock

- Click on the **Stock Manager** tool from the **SOLIDWORKS CAM CommandManager** in the **Ribbon**. The **Stock Manager PropertyManager** will be displayed.
- Select **FCS** option from the **Co-ordinate System** drop-down in **PropertyManager**.
- Specify **5** mm in the **Z+** edit box and click on the **OK** button from the **PropertyManager**. The stock will be created.

Creating Setup

- Click on the **Mill Setup** tool from the **Setup** drop-down in the **Ribbon**. The **Mill Setup PropertyManager** will be displayed.
- Click in the **Entity** selection box and select top face of model for defining mill direction.
- Select the **Face** check box from the **Features** rollout and select **Finish** option from the related **Strategy** drop-down.
- Click on the **OK** button from the **PropertyManager**. The face feature will be created with the setup.

Extracting and Creating Machinable Features

- Click on the **Extract Machinable Features** tool from the **SOLIDWORKS CAM CommandManager** in the **Ribbon**. The machinable features will be extracted automatically. Note that the features created for this practical are 2.5 axis features.
- Select the **Mill Part Setup1** node from the **SOLIDWORKS CAM Feature Tree** and click on the **Generate Operation Plan** button from the **Ribbon**. The machining operations for various features will be created automatically; refer to Figure-31.

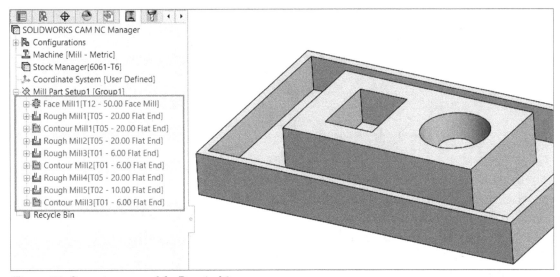

Figure-31. Operations created for Practical 2

Note that the automatic feature has created all the features and operations required to machine the part, so there is no need to manually add any feature or operation.

Generating Toolpaths and Simulation

- Select the **Mill Part Setup1** node from the **SOLIDWORKS CAM Feature Tree** and click on the **Generate Toolpath** tool. The toolpaths will be created for all the operations automatically.
- Click on the **Simulate Toolpath** tool from the **SOLIDWORKS CAM CommandManager** in the **Ribbon**. The **Simulate Toolpath PropertyManager** will be displayed.

- Set the collision buttons to pause in case of collision from the **Options** rollout in **PropertyManager**.
- Click on the **Run** button from the **Navigation** rollout to check the simulation. You will find some toolpaths where collision occurs by tool, shank, and tool holder with stock. Once you reach collision point, click on the **OK** button from the **PropertyManager**. The toolpath in which collision has occurred will be selected automatically.
- Modify the parameters for operations accordingly. You can find the final model with CAM data in resource kit of this book for comparison.

After modifying toolpaths based on simulation, click on the **Post Process** tool from the **SOLIDWORKS CAM CommandManager** in the **Ribbon** and generate the file as discussed earlier.

PRACTICAL 3

Create CNC turning CAM program for model shown in Figure-32. The stock for model is a 105 mm diameter bar of 210 mm length.

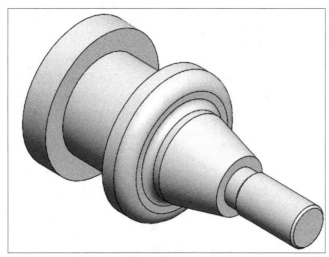

Figure-32. Model for Practical 3

Steps Overview:
1. Start SolidWorks with SolidWorks CAM add-in loaded.
2. Open the part file for Practical 2 of Chapter 8.
3. Define Turning machine and stock.
4. Create the Turning setup and extract the machinable features.
5. Create the features if needed, generate the operations and generate the toolpaths.
6. Check the simulation and post process.

These steps are discussed next in detail.

Starting SolidWorks and Loading CAM add-in

- Double-click on SolidWorks icon from desktop or **Start** menu. The SolidWorks application will start.
- Click on the **Add-Ins** option from the **Settings** drop-down in the **Quick Access Toolbar**. The **Add-Ins** dialog box will be displayed.
- Select the **SOLIDWORKS CAM** check box from the **Active Add-ins** column in the dialog box. If you want to load **SOLIDWORKS CAM** add-in each time SolidWorks starts then select the corresponding check box from **Start Up** column.

• Click on the **OK** button from the dialog box. The add-in will be loaded.

Loading the File

• Click on the **Open** button from the **Quick Access Toolbar** or press **CTRL+O** from keyboard. The **Open** dialog box will be displayed.
• Select the part file C08 Practical 3 from the dialog box and click on the **Open** button. The model file will be displayed.
• Using the **Measure** tool, check various parameters necessary for defining tool type and tool length. For example, for cutting groove in the model, the groove tool must have width less than 10 mm.

Defining Machine and Creating Stock

• Click on the **Define Machine** tool from the **SOLIDWORKS CAM CommandManager** in the **Ribbon**. The **Machine** dialog box will be displayed.
• Select the **Turn Single Turret - Metric** option from the **Available machines** area of **Machine** tab in the dialog box and click on the **Select** button.
• Click on the **Tool Crib** tab in the dialog box and click on the **Add** button to add a grooving tool. The **Tool Select Filter** dialog box will be displayed.
• Select the **Turn Tool** option from the **Tool Type** drop-down at the top in the dialog box. The options will be displayed as shown in Figure-33.

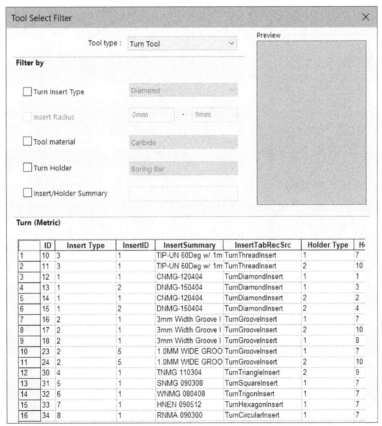

Figure-33. List of turn tools

• Select the 3mm Width Groove Insert tool at ID 16 in the table and click on the **OK** button.
• Click on the **Post Processor** tab and select the machine for which you want to generate CNC program. We have selected HAAS ST20 in our case.
• Click on the **OK** button from the dialog box to setup the machine.

Defining Coordinate System and Creating Stock

- Click on the **Coordinate System** tool from the **Ribbon**. The **Main Spindle Coordinate System PropertyManager** will be displayed.
- Select the **Entity** radio button from the **Origin** rollout and click in the selection box.
- Select the outer round edge of model and define Z direction reference as shown in Figure-34.

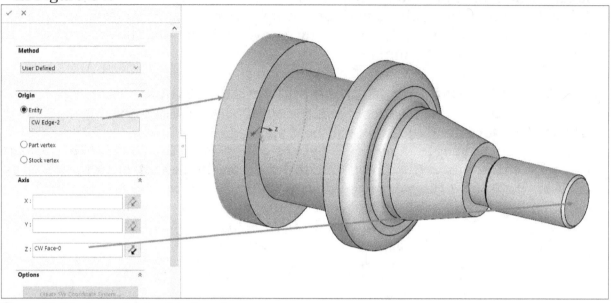

Figure-34. Selection for defining Coordinate System

- Click on the **OK** button from the dialog box to create the coordinate system.

Creating Stock

- Click on the **Stock Manager** tool from the **SOLIDWORKS CAM CommandManager** in the **Ribbon**. The **Stock Manager PropertyManager** will be displayed with preview of stock.
- Specify the diameter as **105** and length of stock bar as **210** in respective edit boxes; refer to Figure-35.

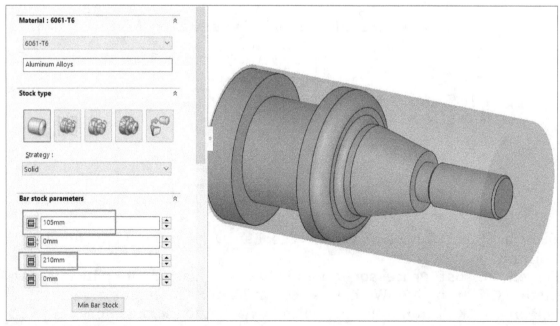

Figure-35. Parameters for stock

• Click on the **OK** button from the **PropertyManager** to create the stock.

Creating Turn Setup and Extracting features

• Click on the **Turn Setup** tool from the **SOLIDWORKS CAM TBM CommandManager** in the **Ribbon**. The **Turn Setup PropertyManager** will be displayed.
• Click on the **OK** button to create turn setup.
• Click on the **Extract Machinable Features** tool from the **Ribbon**. The machinable features will be create automatically.

Creating Operation Plan and Checking Simulation

• Click on the **Generate Operation Plan** tool from the **Ribbon**. The operations will be create automatically; refer to Figure-36. Note that the chuck jaws interfere with groove and turning operations as cutting tool will collide with the jaws. There are two ways to avoid this situation, increase the length of stock or offset jaws backward. In this case, we will increase the length of stock at location attached with jaws.

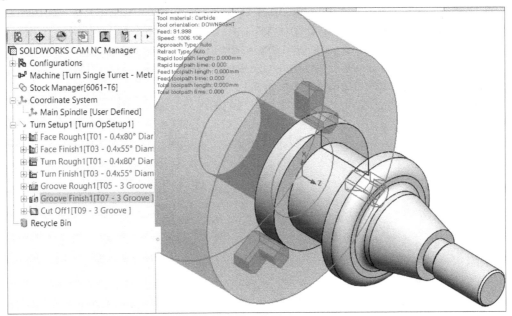

Figure-36. Operations automatically generated

• Click on the **Stock Manager** tool from the **Ribbon**. The **Stock Manager PropertyManager** will be displayed with earlier specified parameters.
• Specify **-35** in **Back Of Stock Absolute** edit box and **240** in **Stock Length** edit box to increase stock enough to avoid accidental collision with chuck jaws; refer to Figure-37.

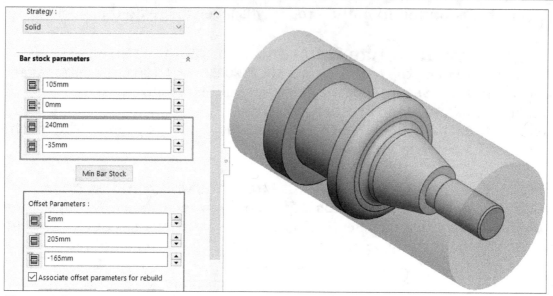

Figure-37. Increasing back stock

- Click on the **OK** button from the **PropertyManager** to modify the stock. The stock will be adjusted automatically and you will be asked to update operations by **SOLIDWORKS CAM Warning** message box. Select the **Yes** button from message box.
- Now, right-click on **Groove Rough** option from the **SOLIDWORKS CAM Operation Tree** and select the **Edit Definition** option from the shortcut menu. The preview of cutting tool will be displayed with operation parameters; refer to Figure-38.

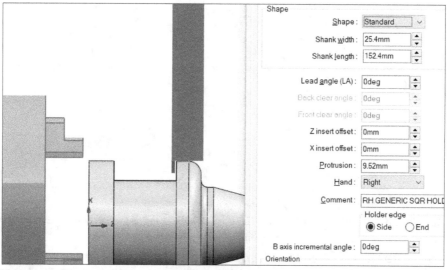

Figure-38. Preview of tool with operation parameters

Note that the length of groove tool is not enough to cut groove and may cause accident with part. So, we need to increase the length of grooving tools.

- Click on the **Groove Insert** tab in the dialog box and increase the length of groove insert by 10 mm which makes length 22.7mm; refer to Figure-39.

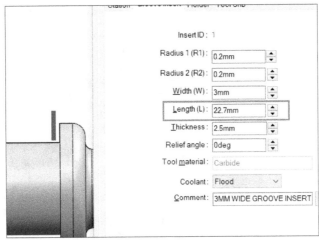

Figure-39. Increasing length of tool

- Click on the **Holder** tab and increase the **Protrusion** value by 10 mm; refer to Figure-40.

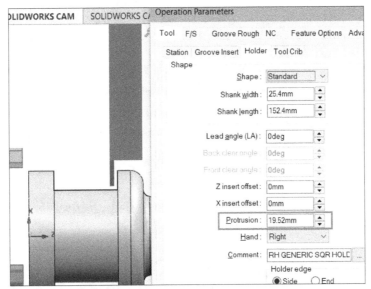

Figure-40. Increasing protrusion of tool

- Click on the **OK** button from the dialog box to modify parameters.

Similarly, modify the parameters for **Groove Finish** operation in **SOLIDWORKS CAM Operation Tree**. After modifying operation parameters, there should be no warning icon left in **SOLIDWORKS CAM Operation Tree**; refer to Figure-41.

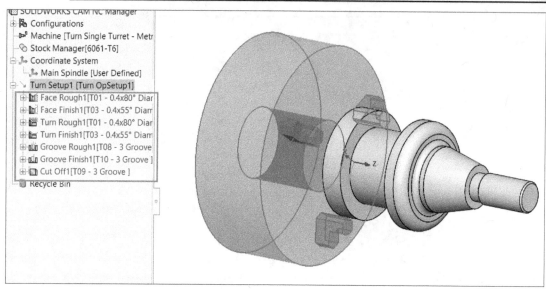

Figure-41. After modifying operations

- Select the **Turn Setup** node from **Operation Tree** and click on the **Generate Toolpath** tool from the **Ribbon**. The toolpaths will be generated for the operations automatically.
- Click on the **Simulate Toolpath** tool from the **Ribbon**. The **Simulate Toolpath PropertyManager** will be displayed with preview of setup.
- Click on the **Run** button from **Navigation** rollout and check the tool movements. Note that while performing the Cut Off operation, there will be a collision between tool holder and stock. You can avoid this collision by increasing the width of cutting tool by 1 mm while keeping width of tool holder as it is.

To generate CNC program codes, click on the **Post Process** button and save the program code file as discussed earlier.

PRACTICE 1

Creating both turning and milling programs for the part given in Figure-42. Stock for the part is a bounding cylinder. Note that you need to create two machining setups for this part.

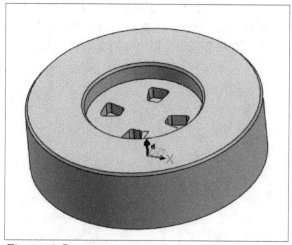

Figure-42. Practice 1

PRACTICE 2

Create NC program for both sides of the model shown in Figure-43. The dimensions of stock for model are: **Diameter = 700** mm and **Length of cylinder = 270** mm; refer to Figure-44.

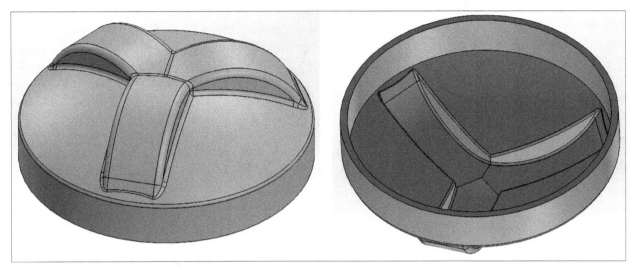

Figure–43. Practice Model

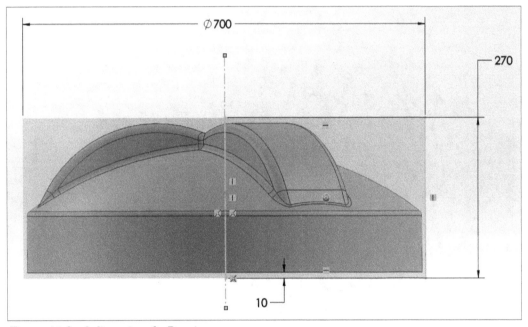

Figure–44. Stock dimensions for Practice

PRACTICE 3

Create NC program for model shown in Figure-45. The stock for model is given in Figure-46.

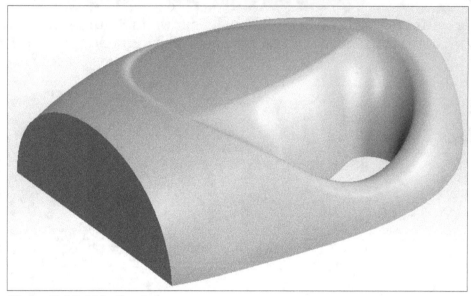

Figure–45. Model for Practical

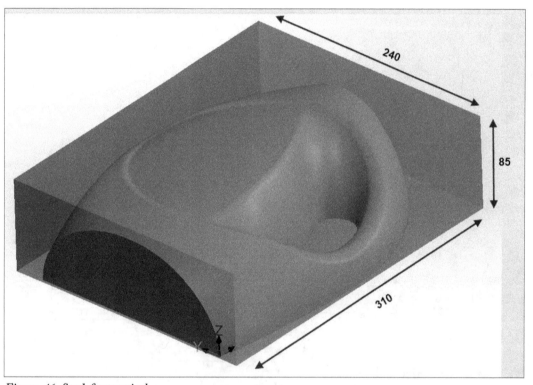

Figure–46. Stock for practical

Index

Ethics of an Engineer

- Engineers shall hold paramount the safety, health and welfare of the public and shall strive to comply with the principles of sustainable development in the performance of their professional duties.

- Engineers shall perform services only in areas of their competence.

- Engineers shall issue public statements only in an objective and truthful manner.

- Engineers shall act in professional manners for each employer or client as faithful agents or trustees, and shall avoid conflicts of interest.

- Engineers shall build their professional reputation on the merit of their services and shall not compete unfairly with others.

- Engineers shall act in such a manner as to uphold and enhance the honor, integrity, and dignity of the engineering profession and shall act with zero-tolerance for bribery, fraud, and corruption.

- Engineers shall continue their professional development throughout their careers, and shall provide opportunities for the professional development of those engineers under their supervision.

OTHER BOOKS BY CADCAMCAE WORKS

Autodesk Revit 2023 Black Book
Autodesk Revit 2022 Black Book
Autodesk Revit 2021 Black Book

Autodesk Inventor 2023 Black Book
Autodesk Inventor 2022 Black Book

Autodesk Fusion 360 Black Book (V2.0.12670)

AutoCAD Electrical 2023 Black Book
AutoCAD Electrical 2022 Black Book
AutoCAD Electrical 2021 Black Book

SolidWorks 2023 Black Book
SolidWorks 2022 Black Book
SolidWorks 2021 Black Book

SolidWorks Simulation 2023 Black Book
SolidWorks Simulation 2022 Black Book
SolidWorks Simulation 2021 Black Book

SolidWorks Flow Simulation 2023 Black Book
SolidWorks Flow Simulation 2022 Black Book
SolidWorks Flow Simulation 2021 Black Book

SolidWorks CAM 2023 Black Book
SolidWorks CAM 2022 Black Book
SolidWorks CAM 2021 Black Book

SolidWorks Electrical 2022 Black Book
SolidWorks Electrical 2021 Black Book
SolidWorks Electrical 2020 Black Book

SolidWorks Workbook 2022

Mastercam 2023 for SolidWorks Black Book
Mastercam 2022 for SolidWorks Black Book
Mastercam 2017 for SolidWorks Black Book

Mastercam 2023 Black Book
Mastercam 2022 Black Book

Creo Parametric 9.0 Black Book
Creo Parametric 8.0 Black Book
Creo Parametric 7.0 Black Book

Creo Manufacturing 9.0 Black Book

Creo Manufacturing 4.0 Black Book

ETABS V20 Black Book
ETABS V19 Black Book
ETABS V18 Black Book

Basics of Autodesk Inventor Nastran 2022
Basics of Autodesk Inventor Nastran 2020

Autodesk CFD 2023 Black Book
Autodesk CFD 2021 Black Book
Autodesk CFD 2018 Black Book

FreeCAD 0.20 Black Book
FreeCAD 0.19 Black Book
FreeCAD 0.18 Black Book

www.ingramcontent.com/pod-product-compliance
Lightning Source LLC
LaVergne TN
LVHW081657050326
832903LV00026B/1790